Collins
Revision

NEW GCSE

Physics

OCR

Twenty First
Century Science

Authors: Michael Brimicombe
Nathan Goodman
Sarah Mansel

Revision Guide +
Exam Practice Workbook

Contents

Contents

Our solar system and the stars

The solar system

- The centre of our **solar system** is a star called the Sun.

- The eight **planets** in our solar system are spherical and have nearly circular **orbits** around the Sun. The four planets closest to the Sun are solid rock; the four outer planets are gas giants.

- Each planet may have a number of **moons** (balls of rock) in near-circular orbit around it.

- **Asteroids** are irregular lumps of rock, mostly in near-circular orbit between Mars and Jupiter.

- **Comets** are small objects made of rock and ice, with very elongated orbits around the Sun.

- **Dwarf planets**, such as Pluto, are small spherical lumps of rock in orbit around the Sun.

- Nearly all (99%) of the solar system's mass is in the Sun.

- Jupiter is the heaviest planet, followed by Saturn, Uranus, Neptune, Earth, Venus, Mars and Mercury.

- Neptune has the largest orbit, followed by Uranus, Saturn, Jupiter, Mars, Earth, Venus and Mercury.

The Universe: distances and sizes

- The Sun is one of thousands of millions of stars in the **Milky Way**.

- The Milky Way is one of thousands of millions of **galaxies** which make up the **Universe**.

diameter of the Milky Way
(100 000 light-years)
25 000 times larger

diameter of solar system
(0.006 light-years)
200 times larger

Earth's orbit
(300 000 000 km)
200 times larger

Sun's diameter
(1 400 000 km)
100 times larger

Earth's diameter
(13 000 km)

distance to the
next nearest star
(4.2 light-years)
700 times larger

distance to the
nearest galaxy
(2 300 000 light-years)
23 times larger

1 light-year = 95 000 000 000 000 km

Comparing sizes and distances
in the Universe.

- Distances to objects outside the solar system are measured in **light-years**. A light-year is the distance that light travels in a year. One light year is 9.5 million million kilometres.

- Light travels through space (a vacuum) at a speed of 3.0×10^5 km/s (300 000 km/s).

- Light takes a very long time to reach us from the stars. We can only observe what stars were like in the past, when the light left them.

EXAM TIP

It is easier to deal with large numbers if they are in standard form. So use 3.0×10^5 km/s instead of 300 000 km/s when you do calculations with the speed of light.

Finding out about the distance to stars

- All the evidence we have about distant stars and galaxies comes from the **radiation** which astronomers can detect.

- Two stars which have the same **real brightness** (appear as bright as each other) can have different **relative brightness**. The star which is further away has a smaller relative brightness.

- If you know the distance to one of the stars, the difference in their relative brightness can be used to calculate the distance to the other one.

- There are uncertainties with this method of measuring the distance to stars:
  - It is based on the assumption that similar types of stars have the same real brightness.
  - It is based on estimating the distance to one of the stars.
  - Many things can make it difficult to make precise observations of stars at night. These include dust, rain, clouds and **light pollution** from streets and buildings.

- Telescopes in space take measurements without distortions from the Earth's atmosphere.

- As the Earth orbits the Sun, nearby stars move slightly against the fixed background of distant stars. This is called the **parallax** effect. Although small, it can be used to find the distance of a star.

- Only stars nearby have a parallax effect which is large enough to be measured.

Improve your grade

The solar system

Foundation: Describe the motion of moons and planets in the solar system.

AO1 [4 marks]

P1 The Earth in the Universe

The fate of the stars

Fusion of elements in stars

- The Sun's energy comes from hydrogen. Hydrogen nuclei are jammed together so hard that they combine in pairs to form the element helium. This process releases loads of energy and is called **nuclear fusion**.

- Fusion in stars forces hydrogen nuclei together to make other elements as well. These other elements spread through space when a star explodes at the end of its life.

- All chemical elements with atoms heavier than helium were made in stars.

- Nuclear fusion is only possible when there are very high densities and temperatures.

- At high enough densities, nuclear fusion can make heavier elements up to iron.

- Heavy stars end their lives as a **supernova**. This is a massive explosion where all the different chemical elements, including those heavier than iron, are made.

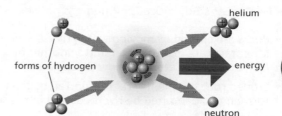

forms of hydrogen

helium

energy

neutron

The fusion reaction that takes place in the Sun.

- The solar system was made from a collapsing cloud of dust and gas about 5000 million years ago.

- Apart from the hydrogen, all of the material in that cloud came from the explosions of large stars. Evidence for this comes from elements in the Sun other than hydrogen and helium.

Remember!

Nuclear fusion is when nuclei are crushed together to form new elements, releasing energy. Nuclear fission breaks up heavy nuclei into lighter ones, releasing energy.

The expanding Universe

- Most of the galaxies appear to be moving away from us.

- This motion of the galaxies increases the wavelength of the light we receive from them.

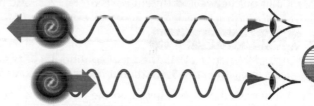

The wavelength of light waves from a star moving away from or towards you changes.

- The increase of wavelength from a galaxy moving away is called **redshift**.

- Apart from a few close galaxies, the amount of redshift increases with distance from Earth. In general, the further away from Earth a galaxy is, the faster it is moving away from us.

- The redshift in the light coming from distant galaxies provides evidence that all parts of the Universe are expanding, with galaxies moving apart from each other as time goes on.

Higher

The Big Bang

- The Universe started expanding rapidly from a single point about 14 000 million years ago.

- The Sun was created about 5000 million years ago.

- The Earth was created about 4500 million years ago.

- Scientists believe that the Universe began with a 'Big Bang'.

- The detection of cosmic background radiation provides evidence to support the **Big Bang theory**.

The future

- The ultimate fate of the Universe depends on how it continues to expand. If there is enough mass in the Universe, gravity will slow down the expansion and make it collapse again.

- The fate of the Universe is difficult to predict because:
  - we can only measure the mass of those parts of the Universe which emit radiation
  - precise measurements of the speed and distance of galaxies is difficult because their radiation travels such a long way to get to us.

Improve your grade

Fusion of elements in stars

Higher: Explain how most of the material in and around you was created by stars.   *AO1* [4 marks]

Earth's changing surface

The changing Earth: erosion and sedimentation

- The surface of the Earth is always changing.
- Material **erodes** slowly from mountains and becomes **sediments** which make rocks.
- **Volcanoes** erupt quickly, spewing out **lava** to make new mountains or a crater.
- Sometimes plants and animals are buried by sediments or lava to become fossils.
- Slow movements of the crust can make fold rocks and new mountains.
- **Geologists** study rocks. Their findings provide evidence of how the Earth has changed.

- Rocks are eroded by moving water, glaciers, wind and rockfalls.
  Mountains are made smaller and smoother by erosion. Valleys are
  made deeper by erosion of riverbed rocks.
- Eroded rock fragments are transported by the wind, water and ice, broken up
  further, and deposited on riverbeds and in the sea. This is called **sedimentation**.
- Over millions of years the sediments are crushed together to form layers of **sedimentary rocks**.
- The build-up of sedimentary rock layers eventually makes seas shallower.

> **Remember!**
> The processes that made and destroyed rocks in the past are still going on today.

- The age of the Earth can be estimated from, and must be greater than, the age of its oldest rocks,
  which are 4000 million years old.
- If no new rocks had been created, erosion for that length of time would have worn all of the
  continents down to sea level.
- Breaks in the Earth's **crust** allow molten rock to escape from volcanoes and create new mountains.
- Collisions between different parts of the crust also push rocks up to make new mountains.

Continental drift

- Alfred Wegener's 1915 theory of **continental drift** says that millions of years ago there was a single
  land mass on Earth. Since then it has split into several continents which have drifted apart.
- Wegener's theory was based on the following evidence:
  - the way continents fit together so well
  - similar fossils and rocks are found on continents now separated by oceans.
- Collisions between moving continents also explains the folding of rocks into mountains.

- Geologists are scientists who study the Earth. They did not accept Wegener's theory because:
  - they already had other, simpler theories which explained some of his observations
  - nobody could explain or measure the movement of the continents
  - Wegener was not a trained geologist and his theory was very different from the others.

- Continents move because they sit on the mantle, whose rocks move slowly by convection as they carry
  heat away from the Earth's hot core.
- The seafloor between continents moving apart can increase by a few centimetres each year. This is called
  seafloor spreading.
- **Higher** • **Oceanic ridges** form on the expanding seafloor where liquid rock from the mantle fills the gap.
- The solidifying rock in oceanic ridges is magnetised by the Earth's field.
- The Earth's magnetic field changes direction over millions of years.
- Each time the Earth's field reverses, so does the magnetisation of the oceanic ridges. So the seafloor has
  strips of reversed magnetism parallel to the gap where new rock is created.

Ideas about science

You should be able to:
- suggest plausible reasons why scientists disagreed with Wegener's theory
- understand that data from experiments are only valid if they can be obtained by other people.

Improve your grade

Continental drift

Foundation: Explain why Wegener's theory of continental drift was not accepted when it was
first published.   *AO1* [4 marks]

Tectonic plates and seismic waves

Tectonic plates

- The Earth's crust has solid **tectonic plates** which float on semi-solid rocks.
- Tectonic plates meet at a **plate boundary**. Here, earthquakes, volcanoes and mountains are found.

- Volcanoes occur when liquid **magma** is forced through cracks where tectonic plates are moving apart.
- Volcanic mountains form when one tectonic plate is forced under another heading towards it.
- **Fold mountains** form when two tectonic plates meet head-on.
- Earthquakes are releases of energy from tectonic plates sliding past each other.
- The **rock cycle** can be explained by the movement of tectonic plates.

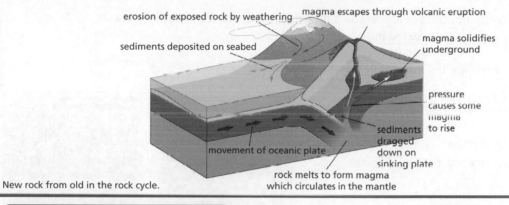

erosion of exposed rock by weathering
magma escapes through volcanic eruption
sediments deposited on seabed
magma solidifies underground
pressure causes some magma to rise
movement of oceanic plate
sediments dragged down on sinking plate
rock melts to form magma which circulates in the mantle

New rock from old in the rock cycle.

Higher

The structure of the Earth

- The core of the Earth is mostly liquid iron.
- Semi-liquid rock in the **mantle** floats on the core.
- The outer core is a layer of liquid nickel and iron about 2200 km thick.
- The inner core is solid nickel and iron about 1250 km thick.
- A thin layer of solid rock in the crust floats on the mantle.

The structure of the Earth.

Seismic waves

- Two types of **seismic waves** are generated when tectonic plates suddenly move.
  - **P-waves** move quickly through solid crust and liquid core.
  - **S-waves** only move slowly through the solid crust.
- Seismometers on the Earth's surface record these waves after an earthquake.

- We can work out the structure of the Earth by measuring the time of arrival of seismic waves across the Earth from an earthquake.
- Seismic waves speed up and change direction when they enter denser regions of the Earth's core.
- P-waves are **longitudinal**, which means the particles vibrate along the direction of motion of the wave.
- S-waves are **transverse**, so the particles vibrate at right angles to the direction of motion of the wave.
- The core of the Earth must be liquid because only P-waves pass through it.

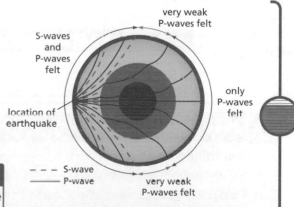

very weak P-waves felt
S-waves and P-waves felt
location of earthquake
only P-waves felt
- - - S-wave
— P-wave
very weak P-waves felt

The timing and strength of detected seismic waves gives information about the Earth's interior.

> **EXAM TIP**
> Make sure that you can clearly describe the difference between a longitudinal wave and a transverse one.

Improve your grade

Tectonic plates

Higher: Explain why there are volcanoes at plate boundaries.

AO1 [4 marks]

Waves and their properties

Finding out about waves

- A **wave** transfers energy away from a vibrating source. The wave creates a series of disturbances as it moves, vibrating material that it passes through.

- There is no overall transfer of matter in the direction of motion of the wave, just energy.

- The **amplitude** of a wave is the maximum height of the disturbance from the undisturbed position.

- The **wavelength** is the distance from one maximum disturbance to the next.

- The speed of a wave is how fast each maximum disturbance moves away from the source:

$$\text{wave speed (m/s)} = \frac{\text{distance travelled (m)}}{\text{time taken (s)}}$$

- The **frequency** of a wave is the number of vibrations of the source in one second.

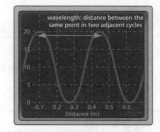

The height of the wave above the surface is the amplitude
The distance between ripples is the wavelength
undisturbed water level

Measuring the amplitude and wavelength of a wave.

- An **oscilloscope** is a machine that displays waves on a screen. A grid on the screen lets you compare the wavelength and amplitude of waves:
  - a sound is louder if it has a larger amplitude
  - a sound is higher pitched if it has a shorter wavelength.

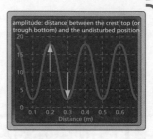

wavelength: distance between the same point in two adjacent cycles

amplitude: distance between the crest top (or trough bottom) and the undisturbed position

Two different sound waves on the same oscilloscope grid. How are they different?

- The scale on an oscilloscope is used to measure wavelengths in metres or to measure amplitude.

- The wavelength and amplitude of the left-hand wave can be found like this:
  - Each horizontal square is 0.1 m. The wavelength is 3.5 squares. So the wavelength is 0.35 m (0.1 m × 3.5 squares).
  - Each vertical square is 5 units. The amplitude is 2 squares. So the amplitude is 10 units (5 units × 2 squares).

Calculating wave frequency and speed

- The unit of frequency is **hertz** (Hz). 1 Hz means 1 vibration per second.

- Waves obey this **wave equation**:

  wave speed (m/s) = frequency (Hz) × wavelength (m).

- The higher the frequency, the shorter the wavelength. The wavelength is always **inversely proportional** to the frequency.

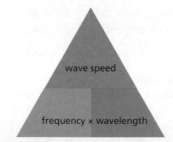

wave speed

frequency × wavelength

Cover up what you want to find out. Use the quantities you know to work out the answer.

EXAM TIP

Always write down the quantities you know with their units, as well as the one you are asked to find. This will help you to select the correct equation.

Ideas about science

You should be able to:

- understand that a scientific explanation can successfully predict the results of experiments

- make predictions from an explanation and compare them with data from experiments to test the explanation.

Improve your grade

Finding out about waves

Higher: Sound in steel has a speed of 2 km/s. What is the wavelength of a sound wave in steel which has a frequency of 80 000 Hz?

AO2 [2 marks]

P1 Summary

Scientists use journals and conferences to inform about their experiments and explanations.

A scientist's judgement about an explanation may be affected by their personal background.

Scientists think up explanations for data from experiments.

The explanations are tested by comparing their predictions with data from new experiments.

Scientific explanations

An explanation is only accepted when it can account for all the data from experiments.

Scientists repeat experiments done by others to check their data.

The solar system was formed from gas and dust in space about 5000 million years ago.

The solar system

The star at the centre of the solar system is called the Sun.

Earth is one of eight planets which orbit around the Sun.

Moons orbit some of the planets.

Comets, asteroids and dwarf planets orbit the Sun.

The distance of a star can be estimated from its brightness and parallax.

The source of the Sun's energy is fusion of hydrogen nuclei, making it hot enough to emit light, which moves through space at 300 000 km/s.

The Sun is one of thousands of millions of stars in the Milky Way galaxy.

The Universe is made up of thousands of millions of galaxies.

Galaxies

The redshift of light from galaxies can be used to measure how fast they appear to be moving away.

Redshift and distance data suggest that the Universe expanded from a single point about 14 000 million years ago.

The expansion of space means that the greater the distance to a galaxy, the faster it appears to be moving away.

The oldest rocks on Earth are about 4000 million years old.

The changing Earth

The Earth's solid crust floats on a soft mantle and is continually worn down by erosion.

Convection currents in the mantle split the crust into tectonic plates and move them.

Mountain building, earthquakes, seafloor spreading and volcanoes happen at the edges of tectonic plates.

The theory of tectonic plates explains Wegener's idea of continental drift.

Earthquakes produce waves which carry energy as they travel through the Earth:

- P-waves are longitudinal waves which can move through the liquid core.
- S-waves are transverse waves which can only move through the solid mantle and crust.

Distance moved by a wave = wave speed × time of travel

Seismic waves

Waves have an amplitude, wavelength, frequency and speed.

Wave speed = frequency × wavelength

Waves which ionise

Electromagnetic radiation

- You see things only when your eyes (detectors) absorb **electromagnetic radiation** called light.
- Some of the things you see are light sources. They emit light, often by glowing hot.
- Everything else that you see **reflects** light. Black objects don't reflect much light, they mostly **absorb** it. Shiny objects reflect most of the light that falls on them.

- Light is one part of the **electromagnetic spectrum**.
- All waves in the electromagnetic spectrum **transmit** through a **vacuum** at 300 000 km/s.
- The energy of a wave in the spectrum increases with increasing frequency. Waves with higher frequencies carry more energy.
- The figure shows how electromagnetic waves are grouped in ranges of frequency.

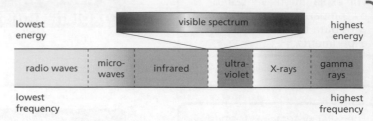

The electromagnetic spectrum. Within visible light, red light has the lowest frequency and violet light has the highest frequency.

Light photons

- Electromagnetic waves transfer energy in packets called **photons**.
- The energy in a photon depends only on the frequency of the wave. Increasing the frequency of an electromagnetic wave increases the energy of its photons.

Radiation intensity

- Solar cells produce electricity. They work by absorbing electromagnetic radiation from the Sun. The solar cell transfers this energy to electrical energy.
- The energy absorbed in each second from an electromagnetic wave depends on its **intensity**. This depends on:
  - the number of photons per second (intensity increases with the number of photons)
  - the energy transferred by each photon (intensity increases with energy).

- The energy of a wave is spread over an increasing area as it moves away from its source.
- This means that the intensity of the wave decreases with increasing distance from its source.

- The intensity of an electromagnetic wave is the energy transferred to each square metre of absorbing surface in each second. The units of intensity are therefore $J / m^2 / s$.
- **Higher:** The intensity of a wave in a vacuum is inversely proportional to the square of its distance from its source.
- If the wave is partially absorbed by the medium it is passing through, the intensity drops even more rapidly than an inverse square law.

Ionisation

- **Atoms** and **molecules** have no overall electric charge. **Electrons** are negatively charged. **Ions** have either positive or negative charge.
- Atoms or molecules are **ionised** when they lose electrons.

- The photons of an **ionising radiation** have enough energy to ionise atoms or molecules.
- The only ionising radiations in the electromagnetic spectrum are high energy ultraviolet, **X-rays** and **gamma rays**.

- Ionisation of a molecule can start off a chemical reaction involving that molecule.

Improve your grade

Ionisation

Higher: Bacteria are single-cell organisms that can pollute drinking water. Explain why exposing the water to ultraviolet light removes the bacteria, but exposing it to visible light does not affect the bacteria.

AO2 [4 marks]

Radiation and life

Effects of ionising radiation

- **Radioactive** materials emit gamma rays. Gamma rays pass easily through the human body.
- X-rays pass through muscle but are absorbed by bone.
- Cells are ionised and damaged when they absorb gamma rays or X-rays. The damaged cells can either die or develop into cancer.

- Physical barriers absorb some ionising radiation.
- X-rays are absorbed by dense materials, so X-rays are used to make shadow pictures of people's bones or their luggage.
- Hospital workers such as **radiographers** are protected from ionising radiation by lead and concrete screens.

Microwaves

- Things which absorb radiation heat up.
- Cells which absorb **radiation** are damaged if they get too hot.
- You can increase the thermal energy transferred from radiation by increasing its exposure time and intensity.

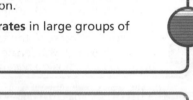
microwaves produced in microwave oven

cold water   hot water

Absorbed radiation has a heating effect.

- Water molecules are good at transferring energy in **microwaves** to thermal energy.
- As the water absorbs the microwaves they vibrate more and share this energy with other molecules around them.
- Cells contain water so heat up when they absorb microwaves.
- Microwave ovens can only be used to heat up food which contains water. Microwaves reflect off the metal in the walls and door or are absorbed by them instead of escaping. This protects people from the radiation.

- Mobile phone networks use low intensity microwaves. The heating effect of these microwaves is very small, but some people are concerned about the health risks from the radiation.
- The risk of cell damage from mobile phones is measured by comparing cancer **rates** in large groups of people who do and do not use mobile phones.
- There is no evidence of harm from mobile phone use.

Ozone

- Sunlight contains ultraviolet radiation. Ultraviolet radiation causes sunburn and skin cancer.
- Sunscreens and clothing protect people by absorbing ultraviolet radiation emitted by the Sun.
- A layer of **ozone** at the top of the atmosphere absorbs ultraviolet radiation from the Sun. This protects living things on Earth from some of the harmful effects of ultraviolet radiation.

- Increased exposure to ultraviolet radiation increases the risk of developing skin cancer.
- If the ozone layer thins by 1%, the risk of skin cancer increases by about 4%.

- Ultraviolet radiation from the Sun causes chemical changes to molecules in the ozone layer.

Higher

Ideas about science

You should be able to:

- understand the possibility of the risk of harm from science-based technologies
- argue that the consequences of a risk, such as skin cancer, need to be considered as well as the chances of it happening
- understand why scientists use large matched samples to investigate correlations between an outcome and a factor
- understand why people underestimate the risk of activities they enjoy, such as sunbathing, and overestimate the risk from unfamiliar or invisible sources, such as ultraviolet radiation.

Improve your grade

Microwaves

Higher: Explain the risks of cooking food with microwaves.   *AO2* [3 marks]

Climate and carbon control

The greenhouse effect

- Radiation from the Sun contains a range of frequencies. Only some of those frequencies pass through the **atmosphere** of the Earth.
- The Earth warms up when it absorbs radiation from the Sun.
- The infrared radiation emitted by the Earth makes it cool down. Radiation from the Earth may pass into space, reflect off clouds or radiate back from gases which absorb it.
- When the Earth's radiation is absorbed or reflected back, this keeps the Earth warmer. We call this the **greenhouse effect**.

Electromagnetic radiation frequencies

- Everything emits some electromagnetic radiation with a range of frequencies.
- The **principal frequency** of that radiation is the one with greatest intensity.
- The principal frequency increases with increasing temperature.
- Radiation from the hot Sun has a higher frequency than radiation from the cool Earth.

Carbon dioxide in the atmosphere

- **Carbon** is found in all living things. It is constantly being recycled through the carbon cycle.
- **Carbon dioxide** in the atmosphere is found in very small amounts. It is absorbed by plants through **photosynthesis** and released by living organisms as they rot (decompose) and **respire**.
- Carbon dioxide is one of the main **greenhouse gases** found in the Earth's atmosphere.
- The level of carbon dioxide in the atmosphere has been steady for thousands of years, because the rates of absorption and release of carbon dioxide have been the same.
- In the last 200 years the level of carbon dioxide has steadily increased because:
  - the burning of **fossil fuels** as an energy source has increased the rate of release
  - cutting down forests (**deforestation**) to clear land has decreased the rate of absorption.
- The recent increase in the temperature of the Earth **correlates** with the rise in carbon dioxide levels in the Earth's atmosphere. Many scientists believe that this correlation is caused by the carbon dioxide because it is a greenhouse gas.

> **Remember!**
> Fossil fuels contain carbon removed from the atmosphere millions of years ago by organisms which were buried before they could decompose.

Global warming

- The greenhouse effect is slowly increasing the average temperature worldwide. This is called **global warming**.
- Global warming will have many effects, including: changes to the crops which can grow in a region; flooding of low-lying land due to rises of sea level as the sea expands and glaciers melt.
- Various gases in the atmosphere are responsible for the greenhouse effect:
  - Water vapour has the most effect, because there is so much of it.
  - The small amount of carbon dioxide has an effect.
  - **Methane** is a strong absorber of infrared radiation, but there is very little of it.
- Scientists use computer models to predict the effects of global warming. The models are tested by seeing if they can use past **data** to predict today's climate.
- Models suggest that global warming will result in more extreme weather events because of:
  - increased water vapour in the hotter atmosphere
  - increased **convection** in the atmosphere increasing wind speed.

Ideas about science

You should be able to:

- understand why correlation, for example between raised carbon dioxide levels and global warming, is not accepted as cause and effect until an explanation has been found.

Improve your grade

Global warming

Foundation: Explain how increasing carbon dioxide in the atmosphere results in global warming.

AO2 [4 marks]

Digital communication

Electromagnetic waves for communication

- Some electromagnetic waves can carry information from one place to another. That information includes text, voice, music and pictures.

- Radio waves and microwaves are not absorbed by air. This means that they can carry radio and TV broadcasts through the atmosphere.

- Infrared and light are not easily absorbed by the glass of **optical fibres**. This means that they can be used for long-distance telephone and internet communication.

- Radio waves use a **carrier wave** to transfer information. The information is used to change the amplitude or frequency of the carrier, in a process called **modulation**.

- A radio receiver demodulates the carrier wave to recover the information.

- An **analogue signal** varies continuously with any value. Sound is an example of an analogue signal.

- Modulation which varies the amplitude of a radio wave continuously makes analogue signals.

Digital signals

- A **digital signal** has only a few values. The digital code for sending sound or pictures has only two values, called 1 and 0.

- By contrast, analogue signals are those which vary continuously.

- Analogue signals, such as sound, can be sent by radio wave in a digital code:
  - The value of the **signal** is measured.
  - The value is coded as a string of 1s and 0s.
  - The 1s and 0s make a digital signal by pulsing the carrier wave on and off (where 0 = no pulse and 1 = pulse).
  - The process is repeated many times a second.

- Radio receivers use the strings of 1s and 0s in the digital signal to recover (**decode**) the original analogue signal and produce a copy of it.

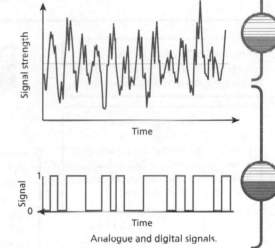
Analogue and digital signals.

- Radio waves can be affected by unwanted information called **noise**. Noise changes the amplitude of the carrier continuously.

- Noise always reduces the quality of the signal recovered by an **analogue** receiver, because the original signal cannot be separated from the noise.

- The wave **pulses** which carry information can usually be separated from the continuously varying noise in a digital receiver, so the quality of the signal is not affected.

Storing digital information

- Digital information is stored as **bytes**.

- A byte is a string of eight **binary digits**. Each binary digit can be either 1 or 0.

- Digital information builds up images from many small dots called **pixels**. The colour and brightness of each pixel is set by binary digits.

- Increasing the number of binary digits for a picture increases the sharpness of the image.

- Digital information builds up sound from a rapid series of values called **samples**. The value of each sample is set by binary digits.

- Increasing the number of binary digits for a sound increases its quality.

- An advantage of using digital signals is that they are easily stored in electronic memories so can be processed by computers.

Improve your grade

Analogue and digital

Foundation: Describe the difference between analogue and digital signals used for radio broadcasts.

AO1 [4 marks]

P2 Summary

Cause and effect

If changes to a factor always result in changes to an outcome, then they are correlated.

Scientists establish correlation by comparing the outcome of large samples whose only difference is the factor.

Scientists will only accept that a factor causes an outcome if they can explain it with theories.

Risk

People's perceptions about risk may be unrealistic: they fear the unfamiliar, but will minimise risks from things they enjoy doing.

We have to make decisions about risk all the time – some risks are greater than others.

Risk can be quantified by measuring the chance of it happening in a large sample over a period of time.

Both the chance of a risk happening and its consequences are considered when making decisions about risk.

Electromagnetic radiation

The intensity of radiation is the energy it delivers per second per square metre of an absorber.

Water in our cells transfers energy from microwave photons to heat. The heating effect depends on both the intensity of the radiation and the length of exposure.

The risk of harm from microwave radiation can be reduced by surrounding sources with metal reflectors.

Low intensity microwaves from mobile phones may be a health risk.

Ultraviolet, X-ray and gamma photons have enough energy to ionise atoms which absorb them.

Cells exposed to ionising radiation can die or become cancerous.

Dense materials absorb X-rays and gamma rays. Sun screens, clothing and the ozone layer absorb ultraviolet radiation.

Electromagnetic radiation carries photons of energy at 300 000 km/s through empty space.

The seven types of electromagnetic radiation in order of increasing photon energy are: radio waves, microwaves, infrared, visible, ultraviolet, X-rays, gamma rays.

The energy of a photon increases with increasing frequency or decreasing wavelength.

Global warming

The Sun transfers energy through the Earth's atmosphere as short wavelength radiation (light). The light is absorbed by the Earth's surface, heating it up.

Burning fossil fuels has increased CO_2 in the atmosphere, causing climate change.

The warmed Earth emits long wavelength radiation (infrared), some of which is absorbed by greenhouse gases (carbon dioxide, water vapour and methane) in the atmosphere, heating it up.

Communication

Radio waves and microwaves carry TV and radio broadcasts because they pass through the atmosphere.

Sound and images can be transmitted as digital signals by switching waves on and off.

The quality of a received sound or image increases with the number of bytes of information transmitted.

Light and infrared carry information long distances along optical fibres.

Although signals become weaker and noisier as they travel, the pattern of 1s and 0s can often be restored at the receiver.

A signal is an electromagnetic wave which has been changed by some information.

An analogue signal is the result of a continuous change.

A digital signal is the result of switching the wave between two values called 1 and 0.

Energy sources and power

Energy sources

- A **primary energy source** is used in the form that it is found. Primary energy sources include:
  - **fossil fuels** such as coal, oil and gas
  - **nuclear fuel**, e.g. uranium and plutonium
  - **biofuels** from plants and animals
  - wind, waves and sunlight.
- Primary energy sources can transfer their energy to **secondary energy sources**, e.g. electricity.

- Many different primary energy sources are used to make electricity.
- Most of our electricity is made from burning fossil fuels. Fossil fuels are a non-renewable energy source (one day there will be none left).

- The burning of fossil fuels is increasing the amount of carbon dioxide, a **greenhouse gas**, in the atmosphere.
- The greenhouse effect is causing **global warming**, an increase in the temperature of the Earth's surface.
- Global warming is likely to result in climate change, leading to floods and storms.

Power

- **Power** is the amount of energy transferred in one second, i.e. the rate of energy transfer.
- A power of one **watt (W)** transfers one **joule** of energy in one second. This is shown by the equation:
  Energy transferred (in J) = power (in W) × time (in s)
- A joule is a very small amount of energy, so domestic electricity measures energy transfer in **kilowatt-hours (kWh)**. There are 1000 watts in 1 **kilowatt (kW)**
- A kilowatt-hour (kWh) is the energy transferred by a power of 1 kW in an hour. This is shown by the equation:
  Energy transferred (in kWh) = power (in kW) × time (in hours)

- The flow of electricity in a circuit is called **current**. It is measured in **amperes**.
- The **voltage** of a power supply is measured in volts (V).
- Electric current in an appliance transfers energy to it from the power supply. The rate at which this is done is found using the equation:
  Electrical power (in W) = current (in A) × voltage (V)

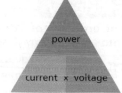

power

current × voltage

Cover up what you want to find out.

- The power of an electrical device is the rate at which it transfers energy from the power supply. This is usually given as a **rating**.
- For an electrical device of a given power rating, e.g. 1000 W, and an electricity supply of a given voltage, e.g. 230 V, the current the device needs to work is found using the equation:

$$\text{Appliance current (in A)} = \frac{\text{appliance power (in W)}}{\text{supply voltage (in V)}}$$

Ideas about science

You should be able to:

- understand that scientists may identify unintended impacts of human activity, such as climate change
- explain how scientists may be able to devise ways of reducing this impact, for example by using renewable fuel sources rather than fossil fuels.

Improve your grade

Power

Foundation: A kettle comes with this warning: *This kettle must be used with a 230 V, 50 Hz supply. The current in the leads will be 8.7 A.*

How much energy in kilowatt-hours is transferred from the supply when the kettle is used for 15 minutes?

AO2, AO3 [3 marks]

Efficient electricity

Buying electricity

- Current in a circuit results in both useful and unwanted energy transfers.
  - Current in a **component** transfers electrical energy into other useful forms.
  - Current in the connecting wires wastefully transfers electrical energy to heat.

- An electricity meter records the amount of electrical energy transferred into a house.
- Electricity is metered in kWh or Units to keep the numbers small and manageable.

- Electricity supplied to a component (in units) = power (kW) × time (in hours)
- Cost of electricity = number of units supplied × cost per unit

> **Remember!**
> Always convert power in W to kW before calculating the number of units used.

Energy diagrams

- Information about energy use can be displayed as:
  - a pie chart to compare different uses of electricity
  - a bar chart to show the amount of electricity used by different components
  - a line graph to show how the number of units consumed changes with time.

- **Sankey diagrams** show the energy transfers in a component. The sum of the energy transfers out of a component equals the input energy. This shows that energy is conserved.

- In the Sankey diagram for a component:
  - Energy flows in from the left. The amount of **energy input** is shown at the left of the arrow tail and is proportional to the thickness of the arrow tail.
  - The sum of the widths of the new arrows at a split is equal to the width of the arrow before the split.
  - Each **energy output** should be shown on the right with an arrow head. Useful energy transfers flow to the right. Wasteful energy transfers flow down.

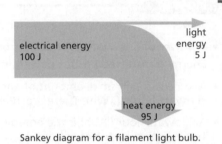

Sankey diagram for a filament light bulb.

Efficiency

- The **efficiency** of a component tells you the proportion of electricity that it transfers into a useful form. It is found using the equation:

$$\text{Efficiency} \times \frac{\text{energy usefully transferred}}{\text{total energy supplied}}$$

- Values for efficiency can range from 1.00 (100%) to 0.00 (0%)
- Components with a high efficiency, e.g. 95%, don't waste much electricity.
- A component with a low efficiency, e.g. 20%, wastes a lot of electricity.

- As a person, you can use less electricity by:
  - using high efficiency A-rated appliances
  - turning off components when they aren't needed
  - not boiling more water than is necessary and cooking food in a microwave oven.
- As a nation, we can use less energy by:
  - using more efficient cars
  - living in houses with better insulation
  - building more efficient power stations and improving the output of old power stations.

- Global demand for energy and other resources will rise in the future as the population increases and the quality of life for many people improves.
- All human activity has an impact on the environment. This impact can be reduced by:
  - recycling resources such as metals, glass and plastics
  - generating electricity from renewable sources of energy, e.g. water, wind and solar power.

Improve your grade

Efficiency

Foundation: Explain why governments have passed legislation which forces people to use energy-efficient lamps in their homes.

AO2 [4 marks]

Generating electricity

Generators

- Moving a magnet near a circuit causes an electric current to flow in the circuit.
- The current flows only when the **magnetic field** is changing – that is, when the magnet is moving.
- Power stations use this idea to produce mains electricity by **generators**.
- A generator contains an **electromagnet** near a coil of wire. There is a voltage across the coil when the electromagnet spins.

Remember!
The magnet has to be spinning near the wire coil for any electricity to be generated – just leaving it there doesn't work.

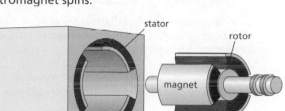

Parts of a generator.

- Power stations use primary fuels, such as fossil, nuclear and biofuels, to boil water into steam.
- The steam passes though a **turbine** (a set of blades on an axle), making its shaft spin round. The turbine shaft spins the magnet inside the generator.
- Increasing the current drawn from the coil requires an increase in the rate of transfer of energy from the primary fuel, i.e. an increase in the amount of primary fuel used each second.
- Primary energy sources used to turn turbines directly include wind and water.
- New developments in electricity generation must conform to government regulations.

How power stations work

- The turbine spins the shaft of the generator to make electricity.
- It is set spinning by steam, hot exhaust gas, wind and water.
- Thermal power stations use coal, oil, gas and nuclear power to spin the turbine from high pressure steam.
  - Coal-fired power stations do this by burning coal to transfer energy into water.
  - Gas-fired power stations burn natural gas to make hot gas for a turbine; another turbine is spun by steam from water heated by the hot gas.
- Hydroelectric power stations use a jet of high pressure water at the base of a dam to spin a turbine.
- Wind-driven power stations use convection currents in the atmosphere caused by the heating effect of the Sun on the land.

| coal burnt in furnace | → | water heated to produce steam | → | jet of steam turns a turbine | → | turbine spins the generator | → | electricity produced by generator |

Processes in a coal-fired power station. Oil-fired power stations are very similar – oil is burned instead of coal to produce heat. They are called 'thermal' power stations.

- Nuclear power stations make high pressure steam by:
  - putting fuel rods close to each other in the **reactor** so that they heat up
    taking away the heat with high pressure water circulating around the rods
  - using the high pressure water to boil low pressure water in a boiler.

Ideas about science

You should be able to:

- understand why the development and application of many areas of science are controlled by governments.

Improve your grade

How power stations work

Foundation: Describe how a power station transfers energy in coal to electricity.   AO1 [5 marks]

Electricity matters

Waste from power stations

- Waste from nuclear power stations is **radioactive** and a serious health risk.
- Nuclear waste must be carefully stored until it becomes safe.

- Nuclear waste emits **ionising radiation**, which affects body cells.
- An object is only **irradiated** when it is placed in the path of the radiation.
- An object is **contaminated** when it gets mixed up with radioactive material.
- Contamination can be a more serious hazard than irradiation because: it can result in a longer exposure to radiation; it is difficult to remove the radioactive material; it is difficult to stop the radioactive material from spreading through the environment.

- We often overestimate the risk from ionising radiation because: it cannot be seen or felt; its effects take a long time to develop; people worry about unfamiliar technology.
- Statistics about death rates can be used to: compare risks from different technologies; decide which technologies need to be controlled; decide which risks are too small to worry about.

Higher

Renewable energy sources

- A renewable energy source can be used over and over again.
- Renewable sources that can spin turbines to make electricity are: **1 hydroelectric** schemes; **2 wind turbines**; **3 wave technology**.

Advantages of hydroelectric schemes	Disadvantages of hydroelectric schemes
• They can provide large amounts of electricity.	• They flood large areas of land.
• They can be turned on and off quickly.	• Rotting plants in the water produce methane gas.
• They can pump water back behind the dam to store energy.	• They cost a lot to build
Advantages of wind turbines	**Disadvantages of wind turbines**
• They are inexpensive to make.	• They need to be put in windy places.
• They need very little maintenance.	• They can only generate electricity when there is enough wind.

- Wave technology is still being developed.

- The environmental impact of renewable energy sources can be reduced by careful planning, e.g. sitting wind turbines offshore, using hydroelectric dams to control rivers.
- The financial impact of building these sources can be offset by the benefits of their use.

The National Grid

- Electricity is convenient because it can transfer energy over long distances for many uses.
- The **National Grid** is a network of cables which carries electricity throughout the UK.

- Power cables warm up because they wastefully transfer electrical current to thermal energy.
- Increasing the voltage of a power cable reduces the current. The National Grid carries electricity at a very high voltage to reduce wasteful energy transfers in the cables.
- Substations connected to the National Grid reduce the voltage to 230 V for our homes.

- There are wasteful transfers of electricity as it is generated and transmitted to the consumer.
- Wasteful energy transfers in transmission are much smaller than those in the power station.
- The efficiency of electricity generation (including the efficiency of the National Grid) is the proportion of the energy in the original fuel that is finally delivered to customers as electricity. In the Sankey diagram opposite it is 33 kJ / 100 kJ = 0.33 (or 33%).

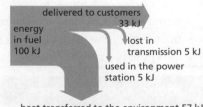

delivered to customers 33 kJ
energy in fuel 100 kJ
lost in transmission 5 kJ
used in the power station 5 kJ
heat transferred to the environment 57 kJ

Where energy is lost in transmission.

Improve your grade

Renewable energy sources

Foundation: The majority of the electricity in the UK is generated from fossil fuels. Explain the advantages and disadvantages of using wind and hydroelectric technology instead. *AO1* [4 marks]

Electricity choices

Choosing the best energy source

- The energy sources that can produce large amounts of electricity are: fossil fuels; nuclear power; hydroelectricity; biofuels.
- The energy sources that rely on the right sort of weather are: wind; waves; solar.
- Energy sources that do not produce greenhouse gases: nuclear power; wind; waves; solar; geothermal.
- Some energy sources have environmental impacts:
  - Fossil fuels produce greenhouse gases; extracting them is dangerous and a pollution risk.
  - Nuclear power creates radioactive waste.
  - Wind farms can cause noise and visual pollution.
  - Hydroelectric and tidal dams flood large areas.
- Some energy sources will eventually run out: fossil fuels; nuclear power.
- Some energy sources are free: wind; hydroelectric; tidal; solar; geothermal.

- When choosing an energy source you should consider:
  - its impact on the environment
  - the cost of building and running the power station
  - how much carbon dioxide and other waste it produces
  - the reliability of the energy source
  - the cost of using the energy source
  - the efficiency of the transfer of energy to electricity.

- The power output of a power station is measured in millions of watts or megawatts (MW).
- A power station which uses nuclear or fossil fuel has a steady output of about 1000 MW and a lifetime of about 40 years.
- Wind farms have a power output of 300 MW, although this varies with the weather, and a lifetime of about 20 years.
- Hydroelectric power stations can have power outputs of about 10 000 MW with lifetimes of about 80 years.

Dealing with future energy demand

- In response to global warming, many countries have agreed to limit their production of carbon dioxide. They can do this by using less energy for transport, heating and electricity.
- Vehicles, factories and power stations emit less carbon dioxide if they become more efficient.
- Switching off appliances at home when they aren't needed reduces energy use.

- Global demand for energy is likely to rise in the future because:
  - there will be more people, especially in developing countries
  - many of them want the high-energy lifestyle of industrialised countries.
- To reduce global energy demand, people in industrialised countries therefore have to use less.
- Industrialised countries already use cleaner, more efficient technology in their workplaces.
- Many goods imported by industrialised countries are made in the developing world with polluting and inefficient technology.

- A secure energy supply in the future requires us to: generate enough electricity to avoid power cuts; have constant access to energy sources; replace old power stations with more efficient ones; use a mix of renewable energy sources; use more renewable energy sources as fossil fuels run out.

Higher

Ideas about science

You should be able to:

- understand that some technologies can impact adversely on the environment and some people's lives, for example by contributing to global warming
- explain that the benefits of using a technology, such as nuclear power stations, should be weighed against the risks, e.g. contamination.

Improve your grade

Dealing with future energy demand

Foundation: It has been suggested that every person on Earth should only be allowed to produce a certain amount of carbon dioxide each year. Explain what impact this could have on your lifestyle. *AO2* [4 marks]

P3 Sustainable energy

People may disagree about which technologies should be allowed. Some decisions may be based on the best outcome for the greatest number of people or a moral sense of right and wrong.

The benefits of science-based technologies must be weighed against their costs.

Human activity may have unintended impact on the environment. Scientists may be able to suggest ways of reducing this impact.

Making decisions

The development of science-based technology is often subject to government regulation.

Many power stations use a primary energy source to boil water into high pressure steam.

The steam passes through turbines, forcing them to spin round. The turbines spin the magnets inside generators so that a voltage appears across the coil inside.

Increasing the rate of energy transfer from the primary source increases the current in the generator.

Electricity can transfer energy easily from one place to another along the National Grid.

Domestic electricity consumption is measured in units of kilowatt-hours (kWh).

Electric current transfers energy from a power supply to an electrical appliance. Energy transferred = power × time

The power of an appliance is the energy transfer per second. Power = voltage × current

Electricity

Electricity is a secondary energy source made from primary energy sources such as fossil fuels, uranium, wind, water and sunshine.

The efficiency of a power station is the ratio of its useful output to the total energy input.

Nuclear power stations produce radioactive waste which emits harmful ionising radiation.

Nuclear power stations are expensive to build but their fuel does not emit carbon dioxide.

Exposure of people to ionising radiation can result in cancer.

Nuclear power

Exposure by contamination is often more dangerous because the radioactive material has been incorporated into the body so cannot be removed.

Exposure by irradiation can be limited by shielding and keeping away from the source of radiation.

Solar cells produce electricity directly from sunlight without producing carbon dioxide.

Solar power requires a lot of land and, like wind power, is very weather dependent.

Hydroelectricity has a large environmental impact as well as being expensive to build.

Geothermal power can only happen in places where rocks are hot near the Earth's surface.

Renewables

Wind turbines are spun directly by the wind. They do not produce carbon dioxide but have a visual impact.

Biofuels, e.g. wood, plant oils, are renewable and result in no net change of carbon dioxide in the atmosphere.

Carbon fuels

Coal, oil and gas produce carbon dioxide when they are burnt to make electricity.

Some fossil fuels will last longer than others, but they will run out eventually.

Fossil fuel power stations can easily make large amounts of electricity but are expensive to build.

Speed

Calculating speed

- The **speed** of an object tells us how far it will travel in a certain time.

- Speed (m/s) = $\dfrac{\text{distance travelled (m)}}{\text{time taken (s)}}$

- Most things do not travel at constant speed, so we usually calculate **average speed**.

- Speed is measured in metres per second (m/s) or kilometres per hour (km/h).

- For example, you walk 3 km to school, and it takes you an hour.
  Speed = 3 km ÷ 1 hour = 3 km/h or: Speed = 3000 m ÷ 3600 s = 0.83 m/s

Remember!
Make sure you convert time into seconds before you calculate speed in metres per second.

- If you measure average speed over a very short time interval you get very close to a value for **instantaneous speed**.

- The **displacement** of an object is the distance from its start point in a straight line. When you run once all the way around a running track, the distance you run is 400 m, but your displacement is zero.

- The displacement is expressed as a distance with a direction.

- Displacement is a vector quantity – it has a size (magnitude) and a direction.

- The **velocity** of an object is its speed in a certain direction. It is also a vector quantity

  Average velocity (m/s) = $\dfrac{\text{displacement (m)}}{\text{time taken (s)}}$

Distance–time graphs

- A **distance–time graph** can be used to visualise a journey.

- The time for the journey is plotted on the *x*-axis (horizontal).

- The distance travelled is plotted on the *y*-axis (vertical).

- A straight line means that the vehicle is travelling at a constant speed. On the graph, line A shows a constant speed.

- A horizontal line means that the vehicle is stationary; speed is zero. On the graph, line B shows that the bus has stopped for 40 seconds after travelling 300 m.

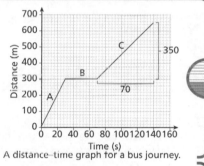

A distance–time graph for a bus journey.

- The **gradient** of the line is equal to the speed.

- The steeper the gradient of the line, the faster the speed. For example, on the graph, line C is less than line A so the bus is travelling slower.

  Gradient of line A = 300 ÷ 30 = 10 m/s
  Gradient of line C = 350 ÷ 70 = 5 m/s

- A curved line means that the speed is changing.

- If the curve is getting steeper it means that the vehicle's speed is increasing.

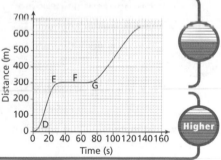

Displacement–time graphs

- Return journeys can be visualised on a **displacement–time graph**.

- When the vehicle has returned to its starting point its displacement will be zero.

- The displacement would be negative if the vehicle travelled *behind* its starting point.

Displacement–time graph: the object moves away at constant speed for 30 s, stops for 40 s and then travels back in the opposite direction for 100 s. It ends up 200 m behind its starting point.

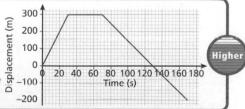

Improve your grade

Calculating speed

Higher: Henry is planning a train journey from Ipswich to Birmingham. The train journey is in two parts:

Departure time		Arrival time		Distance travelled (km)
12:00	Ipswich	13:00	Ely	80
13:15	Ely	15:45	Birmingham	120

a Which train is faster? Show your calculations.

b Taking into account the wait for the connection at Ely, what is the average speed of the journey from Ipswich to Birmingham?

AO1, AO2 [5 marks]

Acceleration

Speeding up and slowing down

- The rate at which the speed of an object increases is known as **acceleration**.
- Acceleration is measured in metres per second squared – m/s^2.

  Acceleration (m/s^2) = $\dfrac{\text{change in speed (m/s)}}{\text{time taken (s)}}$

- So, for a car that speeds up from rest to 25 m/s in 5 seconds, acceleration is: $\dfrac{(25-0)}{5} = 5$ m/s^2

- When an object slows down it has a negative acceleration, sometimes called *deceleration* or *retardation*.
- There has to be net force acting on an object to cause acceleration (or deceleration).
- When the net or overall force is zero, the acceleration is zero.

Changing direction

- A vehicle travelling at constant speed around a corner is changing its velocity.
- Acceleration is defined as the rate of change of velocity.

  Acceleration (m/s^2) = $\dfrac{\text{change in velocity (m/s)}}{\text{time taken (s)}}$

Speed–time graphs

- A **speed–time graph** is used to show the changes in speed during a journey.
- Speed is plotted on the *y*-axis (vertical) and time on the *x*-axis (horizontal).

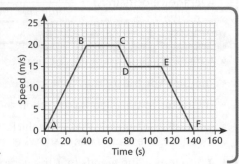

A speed–time graph for a short car journey.

- A horizontal line means that the speed of the object is constant – it is at steady speed. Between points B and C on the graph, the car is travelling at a constant speed of 20 m/s.
- If the horizontal line is along the *x*-axis, the speed is zero – the object is stationary.
- A straight line going up shows acceleration. On the graph, line AB shows acceleration of:
  $(20 - 0) \div 40 = 0.5$ m/s^2
- A straight line going down shows deceleration. On the graph, line EG shows deceleration of:
  $(0 - 15) \div 30 = -0.5$ m/s^2
- The steeper the line, the greater the size of the acceleration.
- The instantaneous speed of a vehicle in a certain direction is its instantaneous velocity.

- A **velocity–time graph** also shows the direction in which an object is travelling.
- A positive velocity means that the object travels in a certain direction and a negative velocity means that the object is travelling in the opposite direction.
- The gradient of the velocity–time graph is equal to the acceleration. For the first 60 seconds, the acceleration of the shuttle train is:
  $(12 - 0) \div 60 = 0.2$ m/s^2

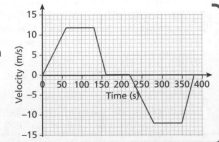

Higher

Improve your grade

Speeding up

Higher: A cyclist rides around a velodrome track. When he sets off he takes 1 minute to reach a top speed of 14 m/s. Then he cycles 10 laps round the oval track at a steady speed.

Calculate the cyclist's initial acceleration, and explain why he continues to accelerate after that.

AO1 [5 marks]

Forces

Forces between objects

- A **force** is a push or a pull which acts between two objects.
- Forces always act in pairs, e.g. you push against a wall; the wall pushes back on your hand.
- A **repulsive** force pushes objects apart; an **attractive** force pulls objects towards each other.

- The size of the force is always the same on both objects.
- The force on each object acts in the opposite direction to each other, e.g. when a ship floats on water, the upthrust (buoyancy) pushes up on the ship and the ship pushes down on the water. This provides a repulsive force pair.
- When you stand you push downwards on the floor. The force is equal to your weight.
- At the same time, the floor exerts an upwards force on you – called the **reaction force** (or the reaction of the surface). It stops gravity from pulling you through the floor.
- The reaction force is equal and opposite to your weight, so you are not accelerating up or down. The reaction force balances your weight.
- If you jump upwards, you need to push harder on the floor – the reaction force increases. Now the forces are unbalanced and you will accelerate upwards.

- Forces are vector quantities, i.e. they have a magnitude (size) and a direction.
- A force is represented by an arrow on a diagram.
- On a scale diagram, the length of the arrow represents the magnitude and the direction of the arrow shows which way the force acts.

The car pushes backwards on the road through the force of friction between the tyres and the road (yellow arrow). The road pushes the car forward with an equal but opposite force (red arrow).

> **Remember!**
> The two forces in an interaction pair are equal in size, opposite in direction and act on different objects.

Rockets (and jets) push gas out of the back of the engine with a large force (yellow arrow). The gas pushes back on the engine with an equal but opposite force (red arrow), propelling the rocket forward.

Friction

- Friction is a force which acts between two surfaces. As the two surfaces slide over one another, friction acts to oppose the motion.
- The size of the friction force depends on:
  - the roughness of the surfaces (rougher surfaces give more friction)
  - how hard the surfaces are pushed together (the heavier the object the more friction).

- When you try to push an object along a surface, friction will be equal to the applied force and acts in the opposite direction, so the object will not move.
- As you increase the applied force, the friction will increase too.
- Eventually the friction reaches a maximum value and the object will start to move – this is called **limiting friction**.
- The kinetic energy of the moving object is transferred to heat energy in both surfaces.
- Lubrication (oil) is used to reduce the friction between moving parts of machinery to stop them getting too hot and wearing out.

- You need friction in order to walk. When you walk your feet push against the friction on the floor, so the floor pushes your foot forwards.
- The **resultant** force (or overall force) acting between your foot and the floor is a combination of friction and the reaction of the surface.
- These forces combine to push the foot in a diagonal upwards direction.

The reaction force (red) and the friction force (blue) combine to push the foot in the direction of the resultant force (black).

Higher

Improve your grade

Forces between objects

Higher: James kicked his football towards the goal. There was a force on the ball when it was kicked. This force was part of an interaction pair.

Describe the partner force of the kicking force in the interaction pair.

AO1, AO2 [3 marks]

The effects of forces

Adding the effects of forces

- There are usually several forces acting on an object at the same time.
- The resultant force is the total or overall force acting on an object.
- When you add forces together you must account for both the size and the direction of the force.

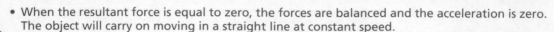

Adding two 5 N forces together can equal 10 N, or 0 N or anything in-between.

- When the resultant force is equal to zero, the forces are balanced and the acceleration is zero. The object will carry on moving in a straight line at constant speed.
- In a frictionless space, once an object starts moving it should keep moving at the same speed.
- When the forces are unbalanced, there is a net force on the object and it will speed up, slow down or change direction.

Terminal velocity

- Objects falling accelerate towards the ground at 9.8 m/s², due to gravity. The force of gravity always acts towards the centre of the Earth.
- The atmosphere creates an upwards force that slows down falling objects. This is known as **air resistance** or **drag**.
- Drag acts in the opposite direction to the speed (or velocity) of the object.
- Drag force increases as the speed of the object increases.
- The larger the surface area of the object, the larger the drag force.

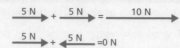

Remember!
In Physics *weight* is the force on an object due to gravity. Weight is measured in Newtons (N), and mass is measured in kg.

- The constant maximum speed reached by a falling object is known as its **terminal velocity**. As the diagram opposite shows:

  A At first the force of gravity is larger than the drag force, so the object accelerates.

  B As speed increases so does drag; the acceleration decreases.

  C When drag equals the force due to gravity there is no resultant force and the acceleration is zero. The object continues at terminal velocity.

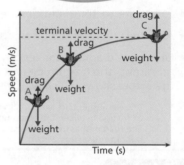

- The ideas about terminal velocity work in the same way for vehicles, with friction and drag acting in the opposite direction to the driving force.

Collisions and momentum

- Large forces are exerted during collisions. The size of the force depends on:
  - the mass of the object (the heavier the object, the larger the force)
  - the speed (or velocity) of the object (the faster the object, the larger the force)
  - the duration of the impact (the longer the time to stop, the lower the force).
- Most car safety devices, e.g. air bags, crumple zones, seat belts, crash helmets, are designed to increase the impact time, thus reducing the force in a collision.

- **Momentum** (kg m/s) = mass (kg) × velocity (m/s)
- A resultant force will change an object's momentum. The larger the force exerted, the larger the change in momentum.

  Change of momentum (kg m/s) = resultant force (N) × time for which it acts (s)

Remember!
Momentum has both direction and magnitude so it is a vector quantity.

- Force is equal to the rate of change of momentum.

  $$\text{Force (N)} = \frac{\text{change in momentum (kg m/s)}}{\text{time taken (s)}}$$

- For example, a car of mass 1200 kg crashes into a wall at a speed of 20 m/s. The collision stops the car in a time of 1.5 s.

  Change in momentum = (1200 × 20) − 0 = 24 000 kg m/s

  Force = 24 000 ÷ 1.5 = 16000 N

Improve your grade

Terminal velocity

Higher: Explain how air bags reduce injury in a car crash. Include ideas about momentum in your answer.

AO1, AO2 [3 marks]

Work and energy

Work

- The energy used by the movement of a force is known as the **work** done.
- Energy and work are both measured in joules (J). Energy is defined as the ability to do work.
- Work done (J) = Force (N) × distance moved in the direction of the force (m)

- When work is done *on* an object, energy is transferred to that object.
- When work is done *by* an object, energy is transferred from the object to something else.
  Amount of energy transferred (J) = Work done (J)

- All forms of energy have the potential to do work.
- Energy from food is transferred in our bodies so we can do exercise.
- Not all energy is transferred as work – some is always dissipated as heat.

Potential energy

- When you lift an object, you do work against gravity.
- 1 joule of work will lift a weight of 1 Newton a distance of 1 metre.
- The work is transferred to **gravitational potential energy** (GPE) of the object.

- As an object is raised, its gravitational potential energy increases. As an object falls, its gravitational potential energy decreases.
- Change in gravitational potential energy (J) = weight (N) x vertical height difference (m)
  For example, if your weight is 700 N and you climb stairs a height of 3 m:
  Gain in gravitational potential energy = weight × height = 700 N × 3 m = 2100 J

Kinetic energy

- When you push an object to get it moving (increase its velocity) you do work. The work is transferred to the moving object as **kinetic energy** (KE).
- The greater the mass of the object and the faster its speed, the greater its kinetic energy.

- Kinetic energy (J) = ½ × mass (kg) × [velocity]² ([m/s]²)
  For example, for a car of mass 800 kg travelling at a speed of 12 m/s:
  KE = ½ × mass × velocity² = ½ × 800 × 12² = 400 × 144 = 57 600 J

> **EXAM TIP**
> Always remember to square just the velocity – do that step first in your calculation.

- The work done by an applied force is the same as the change in the kinetic energy of the object.
  For example, the driving force of a car is 8 kN and it moves a distance of 7.2 m:
  Work done = force × distance moved = 8000 x 7.2 = 57 600 J – change in kinetic energy

Energy transfers

- As a roller coaster travels round its track, going up and down, its energy changes from kinetic energy to gravitational potential energy.
- Its total energy at any time is the sum of its kinetic energy and its gravitational potential energy.
- When there are no resistive forces the total energy remains constant. This is known as the principle of **conservation of energy**.

- Energy can be neither created nor destroyed; it can only transfer between objects or change its form.
- Usually some energy is used up doing work against friction and air resistance – this means that some energy is dissipated as heat.

- When an object falls, its potential energy is transferred to kinetic energy.
- Ignoring energy transferred due to friction and air resistance:
  Loss in gravitational potential energy = gain in kinetic energy

Improve your grade

Energy transfers
Foundation: A spacecraft is returning to Earth. It has a gravitational potential energy of 8 MJ on re-entry.
a What is its maximum possible increase of kinetic energy as it falls?
b Explain why the actual increase of kinetic energy will be less than this value. *AO1, AO2* [3 marks]

P4 Summary

The gradient of a distance–time graph is the speed.

A horizontal line means that the speed is zero.

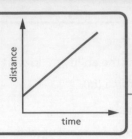

Speed and acceleration

$$\text{Average speed (m/s)} = \frac{\text{distance travelled (m)}}{\text{time (s)}}$$

The gradient of a speed–time graph is the acceleration.

A horizontal line means the speed is zero.

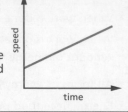

Displacement is the distance travelled in a certain direction.

Velocity is the speed in a certain direction.

Both displacement and velocity are vector quantities.

Acceleration causes a change in speed and/or a change of direction. Retardation is negative acceleration causing a decrease in speed.

$$\text{Acceleration (m/s}^2) = \frac{\text{change in speed (m/s)}}{\text{time (s)}}$$

Forces are vector quantities – they have both size and direction.

When more than one force acts on an object the resultant force is the sum of all the individual forces.

Forces occur in 'interaction pairs' which always:
- are the same size
- act in opposite directions.
- act on different objects.

When the resultant force on an object is zero:
- the forces are balanced.
- there is no acceleration. The object carries on moving in a straight line at constant speed.

Forces and motion

Friction forces act between two surfaces in the opposite direction to motion.

Reaction forces act on objects upwards from surfaces.

When an object is travelling in a straight line, if the driving force is:
- greater than the drag forces, the object will speed up
- equal to the drag forces, the object will continue to move at constant speed – sometimes known at 'terminal velocity'
- smaller than the drag forces, the object will slow down.

Momentum (kg m/s) = mass (kg) × velocity (m/s)

Momentum

To decrease the size of forces in collisions, the time taken to stop can be increased by using: air bags; seat belts; crumple zones; crash helmets.

A resultant force will cause a change in momentum.

$$\text{Force (N)} = \frac{\text{change in momentum (kg m/s)}}{\text{time taken (s)}}$$

When a force moves an object it does work.

Work done (J) = force (N) × distance moved (m)

When work is done energy is transferred.

Work done (J) = energy transferred (J)

The energy of a moving object is its kinetic energy.

The faster the motion and the heavier the object, the larger the kinetic energy.

Kinetic energy (J) = ½ x mass (kg) × [velocity]² ([m/s]²)

Energy is always conserved – it can neither be destroyed nor created, only transferred.

In any energy transfer, some energy is dissipated or wasted as heat.

Work and energy

As you raise an object, its gravitational potential energy (GPE) increases.

GPE (J) = weight (N) × height (m)

Electric current – a flow of what?

Static electricity

- An **atom** is made up of charged particles. It has a positive **nucleus** with negative **electrons** orbiting it. The nucleus is made up of **neutral neutrons** and positive **protons**.

- Neutral objects have no overall charge.

- There are **electrostatic forces** between charged objects, e.g. hair stands up when it is attracted to a charged comb.

- Like charges repel; unlike charges attract.

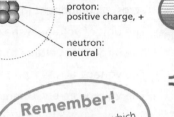

electron: negative charge, –

proton: positive charge, +

neutron: neutral

- There is an electrostatic force of attraction between the positively charged nucleus and the negatively charged electrons in an atom.

- The outermost electrons are less strongly attracted to the nucleus and can be removed by rubbing.

- When two insulating objects are rubbed together they become charged, because electrons are transferred from one object to the other.
  - The object which has lost electrons will become positively charged.
  - The object which has gained electrons will become negatively charged.

> **Remember!**
> It is only the electrons which are transferred when objects are charged by friction.

- When you brush your hair, individual hairs become similarly charged and repel each other, making your hair stick up.

- When you take off a nylon or polyester top, there can be a spark or a crackle over your head. This is caused by the electrons moving through the air from the negatively charged clothing to your positively charged hair.

- During a thunderstorm, charge builds up in the clouds. When the amount of charge becomes large enough to break down the insulation of the air, the charge flows between the cloud and the Earth as a flash of lightning.

Conductors and insulators

- Metals are good electrical **conductors** because they have **free electrons**. This means there are lots of charges free to move.

- Plastics are electrical **insulators**. There are few free electrons in plastics, so there are few charges free to move.

Free electrons in the metal cannot pass through the plastic layer.

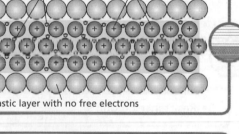

free electrons from outer shells of metal ions   metal ions

plastic layer with no free electrons

Moving charges

- When the bulb is lit in a circuit, there is an **electric current**.

- The moving electrons, or electric current, transfer energy to light the bulb.

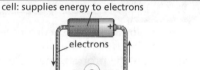

cell: supplies energy to electrons

electrons

electrical energy transferred to light energy

- In a complete circuit, there are free electrons in all the metal components and connecting wires. The cell (or battery) supplies energy to the electrons. The electrons carry a negative charge, so they will flow from the negative terminal of the cell towards the positive terminal.

- The flow of charge is the electric current.

- The bulb is converting the energy carried by the electrons into light (and heat) energy.

- Electric current is the **rate of flow of charge**, or the charge flowing per second.

- Electric current is measured in **Ampères**, or amps (A) for short.

- The more energy the charged particles receive from the power supply, the greater the current.

- In an electric circuit, charge is conserved and energy is transferred.

Improve your grade

Static electricity
Higher: Bella was rubbing a nylon comb with a duster. When she put the comb near her head, her hair moved towards the comb. Explain why this happened.   AO1, AO2 [5 marks]

Current, voltage and resistance

Measuring current and voltage

- An **ammeter** is used to measure current. An ammeter is connected in series. The circuit symbol for an ammeter is: —(A)—

- A **voltmeter** is used to measure **voltage**. A voltmeter is connected in parallel across a **component** in a circuit. The circuit symbol for a voltmeter is: —(V)—

- The unit of voltage is the **volt** (V).

- The larger the voltage of the battery of in a circuit, the bigger the current.

- The voltage across a power supply is a measure of how much energy is supplied to the circuit.

- The voltage across a component is a measure of how much energy is transferred in the component.

> **EXAM TIP**
> Always talk about the current *through* a component and the voltage *across* a component.

- In a circuit, the charges (free electrons) are the energy carriers. They collect energy at the power supply and transfer energy at a component.

- **Power** is the rate at which energy is transferred. It is measured in watts (W).
  Power (W) = voltage (V) × current (A)

- A voltmeter measures the difference in energy between the terminals of a battery or bulb.

- The difference in energy per unit charge is known as the **potential difference (p.d.)**. This is the scientific term for voltage.

- A potential difference of 1 volt means that 1 joule of energy is transferred into or out of electrical form for each unit of charge.

Electrical resistance

- The more **resistance** in a circuit, the lower the current.

- All electrical components, e.g. lamps, motors, have some resistance to the flow of charge through them.

- The greater the voltage across a resistor, the larger the current.

- The circuit symbol for a resistor is: —[ ]—

- Resistance is a measure of how much a conductor opposes the current. Its unit is the ohm (Ω).

- Copper wires have such a low resistance that we can ignore it.

- A **variable resistor** is a device that allows you to control the current by changing the amount of resistance wire in a circuit.

- The symbol for a variable resistor is: —[⟋]—

- Resistance (in ohms) = $\dfrac{\text{voltage (in volts)}}{\text{current (in amps)}}$ or: $R = \dfrac{V}{I}$

Ohm's law

- A graph of voltage against current will give a straight line through the origin. This means that the current through a fixed resistor is directly proportional to the voltage across it (at constant temperature).

- The higher the resistance, the lower the gradient.

> **Remember!**
> Resistance is equal to 1/gradient of the line *only* if the voltage is along the x-axis and current is on the y-axis.

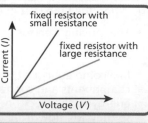

Ideas about science

You should be able to:

- suggest why using experimental values of current or voltage may not give the true value of resistance in a circuit.

Improve your grade

Electrical resistance

Higher: Harry was recording values of current and voltage across a resistor so he could calculate the resistance.

a He recorded a current of 0.06 A when the voltage across the resistor was 4 V. Calculate the resistance.

b Harry then repeated the experiment with a voltage of 6 V. What current reading did he expect to get? Explain why he might not get this exact value.
AO1, AO2 [5 marks]

Useful circuits

Series and parallel circuits

- Components connected **in series** are in a line.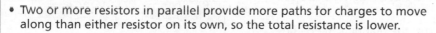
- The current is the same through all the components connected in series.
- The more cells connected in series, the greater the potential difference.
- The potential difference across the components adds up to the p.d. across the battery.
  Supply p.d. = p.d. across R_1 + p.d. across R_2 + p.d. across R_3
- The p.d. across each component will be in proportion to its resistance.
- The overall resistance will be the sum of all the individual resistances: $R_{total} = R_1 + R_2 + R_3$

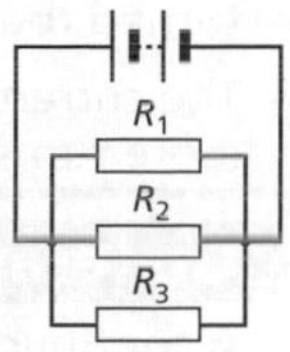

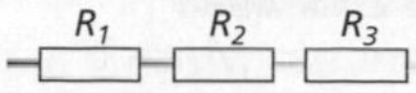

- Components **in parallel** are each connected separately to the power supply (see right).
- The charge has a choice of pathways, so the current is shared between each branch. The current to and from the power supply is the sum of the current through all branches. $I_{total} = I_1 + I_2 + I_3$
- Two or more resistors in parallel provide more paths for charges to move along than either resistor on its own, so the total resistance is lower.
- The current through each resistor is *inversely proportional* to its resistance, i.e. it is largest through the component with the smallest resistance.
- Work is done by the power supply to provide energy to the charged particles. A bulb uses the energy to do work to provide heat and light; a resistor uses the energy to do work to provide heat.

> **Remember!**
> You can use the relationship $R = V/I$ for each branch of the circuit.

- In a series circuit, the work done on each unit of charge by the battery must equal the work done on it by the circuit components.
  - More work is done by the charge moving through a large resistance than through a small one.
  - Two or more resistors have more resistance than one on its own, because the battery has to move charges through both of them.
  - A change in the resistance of one component, e.g. variable resistor, will cause a change in the potential differences across all the components.
- Cells connected in parallel will have the same potential difference as one cell on its own, but the amount of energy in the circuit will increase.
- The potential difference across components connected in parallel is always the same, and is equal to the pd of the battery.

Thermistors and LDRs

- A **thermistor** is a semiconductor whose resistance changes with temperature. 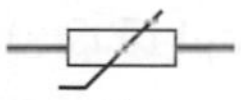
- An **light dependent resistor (LDR)** is a semiconductor whose resistance changes as the amount of light falling on it changes. In bright light the resistance will be low.

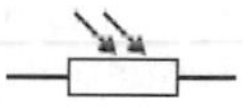

- The left graph shows the variation of resistance with temperature. At high temperature the resistance is lower.
- The right graph shows the variation of resistance with light intensity for an LDR.

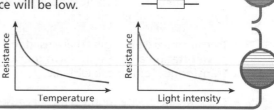

Metals and semiconductors

- The graph opposite shows how the current through a bulb varies with increasing potential difference. As the wire in the bulb gets hotter, its resistance increases.
- In semiconductors, as the temperature or light intensity increases there are more free electrons to carry the current, so the current is higher.

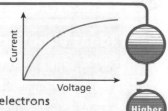

- The positive ions in the metal structure have more energy and vibrate more. The free electrons will collide more often. This means that they cannot move as fast, so the current decreases.

Improve your grade

Series circuits and parallel circuits

Foundation: Susan and Mark were discussing adding resistors to a circuit. Susan said that if you added more resistors the total resistance would increase. Mark told her that sometimes adding more resistors would decrease the total resistance. Explain why both Susan and Mark are correct. *AO1, AO2* [5 marks]

Producing electricity

Making an electric current

- A **magnetic field** is a space around a magnet in which a magnetic force acts.
- The magnetic field is strongest where the field lines are closest together.

- A voltage is **induced** when a magnet is moved near a piece of wire (or when a wire is moved near a magnetic field). If the piece of wire is part of a circuit, a current will flow.
- A voltage is always induced when there is relative movement between a magnet and a coil of wire.
- The direction of the current is reversed when the motion of the wire is reversed, or the magnet is turned round.
- The current will increase if the speed of motion increases, a stronger magnet is used or there are more turns of wire in the coil.

Generators

- A continuous supply of electricity is produced when there is continuous relative motion between a magnet and a coil of wire.
- The coils of wire continuously 'cut' the magnetic field lines so a voltage is induced. This is called **electromagnetic induction**.

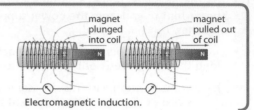
Electromagnetic induction.

- A larger voltage is induced if:
  - the strength of the magnet is increased
  - the number of turns in the coil is increased
  - an iron core is used inside the coil
  - the rate at which the coil is turned is increased.
- The faster the rate of cutting field lines, the larger the induced voltage.
- Mains electricity is produced by generators in power stations that induce an alternating voltage.

- As a coil rotates in a uniform magnetic field it cuts the lines of magnetic field at different rates:
  - When the coil is at right angles to the field lines, it cuts no field lines, so the induced voltage is zero.
  - When the coil is parallel to the field lines, its rate of cutting field lines is at a maximum, so the inducted voltage is at a peak.
  - As the coil rotates, it cuts field lines in a different direction, so the direction of the voltage alternates.

Distributing mains electricity

- **Direct current (d.c.)** always flows in the same direction. Batteries produce d.c. electricity.
- **Alternating current (a.c.)** changes direction at regular intervals.
- In the UK, mains electricity is a.c. and is generated at 230 V and at a frequency of 50 Hz. It is transmitted along cables held up by pylons at very high voltages.

- Transformers (see page 31) are used to step up the voltage at the power station and to step down the voltage near our homes. Transformers only work with a.c.

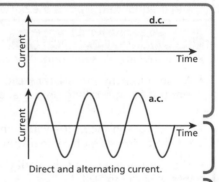
Direct and alternating current.

- The UK's electricity supply is a.c. because it is easier to generate a.c. electricity in large amounts. Also, many different fuels can be used in power stations.
- It is more efficient to transmit electricity at very high voltages.

Ideas about science

You should be able to:
- identify factors that might affect an outcome
- make predictions about the outcome of changing a variable.

Improve your grade

Generators

Foundation: Jenny uses a dynamo to power her bicycle lights.

a Explain why the lights are dim when she cycles slowly.

b Suggest one advantage and one disadvantage to using dynamo lights instead of battery-powered lights.

AO1, AO2 [5 marks]

Electric motors and transformers

The magnetic effect of a current in motors

- **Motors** are used in many electrical appliances that make things move, e.g. hairdryer, DVD player.

- There is a circular magnetic field around a wire carrying electric current.

- If the wire is made into a coil, the magnetic field pattern becomes similar to that of a bar magnet. This is called an **electromagnet**.

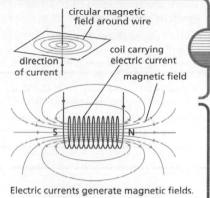

- When a current flows in a wire that is in a magnetic field, the wire experiences a force. If the wire is free to move, it moves. This is called the **motor effect**.

- The force is largest when the current is at right angles to the magnetic field lines. The direction of the force is always at right angles to both the current in the wire and the magnetic field lines.

- No force is experienced when the current is parallel to the magnetic field lines.

- The direction of the force is reversed if either the current or the magnetic field is reversed.

Electric currents generate magnetic fields.

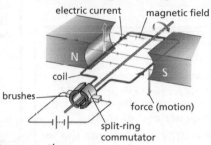

- When a simple motor (see opposite) is placed in a uniform magnetic field, one side of the rectangular current-carrying coil is forced upwards, while the other is forced downwards to produce rotation.

- The motor will turn faster if: the current is increased; the number of turns on the coil is increased; the magnetic field is stronger; there is a soft iron core in the coil.

An electric motor.

- The motor effect works because of the interaction between the magnetic field around the current carrying wire and the magnetic field of the permanent magnets.

- In the diagram, the two magnetic fields reinforce each other above the wire and cancel each other out below the wire, so the wire is forced down.

- The coil of the motor is connected to the power supply using a commutator. The commutator swaps contacts with the coil every half turn to reverse the current through the coil. This keeps the motor turning.

Combining magnetic fields.

Transformers

- A transformer changes the voltage of an a.c. power supply.

- It consists of two separate coils around an iron core.

- The input voltage is fed into the **primary coil**.

- The output voltage is across the **secondary coil**.

- The larger of the two voltages will be across the coil with the most turns.

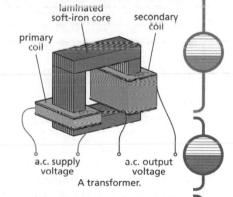

- A **step up transformer** converts a low voltage input to a higher voltage output. The primary coil will have fewer turns than the secondary coil.

- A **step down transformer** converts a high voltage input to a lower voltage output. The primary coil will have more turns than the secondary coil.

A transformer.

- The alternating current in the primary coil creates an alternating magnetic field around it. The magnetic, soft iron core channels the magnetic field through the secondary coil.

- The alternating magnetic field will continuously cut through the wires in the secondary coil and an a alternating voltage will be induced across the secondary coil.

- If the number of turns in the secondary coil is doubled, the output voltage will be doubled.

- The turns ratio is equal to the voltage ratio:

$$\frac{\text{voltage across primary coil}}{\text{voltage across secondary coil}} = \frac{\text{number of turns in primary coil}}{\text{number of turns in secondary coil}}$$

Improve your grade

Transformers

Higher: The mains supply at home is at 230 V a.c. A computer needs a supply at 23 V. Describe how the voltage of the mains is converted to the lower voltage.

AO1 [5 marks]

P5 Summary

Electric charge

There are two types of charge – positive and negative.

Like charges repel; unlike charges attract.

Metals are good electrical conductors because they contain many free electrons.

Plastics are insulators because there are few free electrons.

Electrostatic effects are caused by a transfer of electrons.

A positively charged object has lost some electrons.

A negatively charged object has gained some electrons.

Electric current is the rate of flow of electric charge.

Electric current is measured in amps (A).

Circuit electricity

Energy is supplied to a circuit by a battery or cell.

The voltage or potential difference (p.d.) across the cell is a measure of the amount of energy per unit of charge.

The energy is transferred to the component in the circuit.

An LDR is a semiconductor whose resistance decreases in brighter light.

A thermistor is a semiconductor whose resistance usually decreases as temperature increases.

The resistance of metals increases with increasing temperature.

Resistors oppose the motion of the charged particles, and get hot.

Resistance is measured in ohms (Ω).

The higher the resistance, the lower the current. The higher the voltage, the higher the current.

Ohm's law states:

$$\text{Resistance } (R) = \frac{\text{voltage } (V)}{\text{current } (I)}$$

When resistors are connected in series, the total resistance is increased.

The current around a series circuit is the same everywhere.

The p.d. across the components adds up to the p.d. across the power supply.

$V_{total} = V_1 + V_2 + V_3$

The p.d. is split in proportion to the resistance.

When resistors are connected in parallel, the total resistance is less than any of the resistors individually.

The p.d. across each component is equal to the p.d. across the power supply.

The current has a choice of pathways.

The total current through the power supply is the sum of the currents through each component. $I_{total} = I_1 + I_2 + I_3$

Generating electricity

Alternating current (a.c.) changes direction at regular intervals.

Direct current (d.c.) always flows in the same direction.

Mains electricity supply in the UK is 230 V at 50 Hz a.c.

Electricity generators work by electromagnetic induction.

When a wire moves relative to a magnetic field, a voltage is induced.

Transformers change the voltage of a.c. electricity. Step up transformers increase voltage; step down transformers decrease it. A transformer consists of two coils around an iron core. A varying current in the primary coil produces a varying magnetic field; the varying magnetic field induces a voltage in the secondary coil.

$$\frac{\textit{voltage across primary coil}}{\textit{voltage across secondary coil}} = \frac{\textit{number of turns in primary coil}}{\textit{number of turns in secondary coil}}$$

In a generator, a magnet is rotated within a coil of wire to induce a voltage. The size of the induced voltage can be increased by: increasing the speed of rotation; the strength of the magnetic field; the number of turns on the coil; placing an iron core inside the coil.

Electric motors

Motors are used in all electrical appliances that convert electrical energy to movement, e.g. vacuum cleaners, CD players.

The motor effect:

A wire carrying a current experiences a force when in a magnetic field. The force is at maximum value when the wire is at right angles to the magnetic field lines. Continuous rotation of a wire coil can be produced if the current is reversed in the coil every half turn, with the use of a commutator.

Nuclear radiation

The nuclear atom

- An **atom** is the smallest part of an element.

- **Neutrons** and **protons** form the **nucleus** of an atom. **Electrons orbit** the nucleus at high speed.

- Electrons are the smallest particle and are negatively charged.

- Protons have mass 2000 times that of an electron and are positively charged.

- Neutrons have the same mass as protons and have no charge.

- An atom is neutral, so has an equal number of electrons and protons.

- Nearly all the mass of the atom is concentrated in the nucleus.

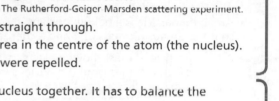

An atom of helium

- Evidence about the structure of the atom was obtained from 1909 Rutherford-Geiger-Marsden alpha scattering experiment during which alpha particles were fired at gold foil.

- The observations recorded were:
  - Most **alpha particles** passed straight through the gold foil undeviated.
  - A few particles were deflected through small angles.
  - Even fewer bounced straight back from the foil.

- The conclusions were:
  - The atom is mostly empty space because most particles went straight through.
  - The mass and charge of the atom is concentrated in a small area in the centre of the atom (the nucleus).
  - The nucleus was positive because the positive alpha particles were repelled.

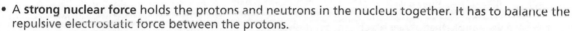

The Rutherford-Geiger Marsden scattering experiment.

- A **strong nuclear force** holds the protons and neutrons in the nucleus together. It has to balance the repulsive electrostatic force between the protons.

- The number of protons in a nucleus determines the element and its chemical properties.

- **Isotopes** are atoms of the same element with the same number of protons but differing numbers of neutrons in the nucleus. For example, the three isotopes of hydrogen are hydrogen (1 proton, 1 neutron), deuterium (1 proton, 2 neutrons) and tritium (1 proton, 3 neutrons).

Higher

Radioactive elements

- **Radioactive** elements are unstable and constantly emit **ionising radiation**. This makes them become more stable.

- Ionising radiation knocks out electrons from atoms, forming a positive **ion**.

- Ionising radiation can be either high-energy particles or high-energy electromagnetic waves.

- **Background radiation** is low-level ionising radiation that is all around us.

- Some background radiation comes from outer space as cosmic rays, but most comes from rocks and soil.

Remember!
Not all background radiation comes from natural sources – some comes from industry and medical uses.

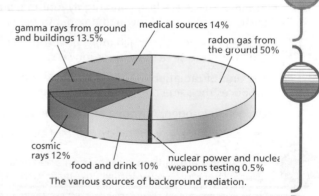

The various sources of background radiation.

- Radioactivity is a random process. We never know when an unstable nucleus will decay and emit ionising radiation.

- The amount of radiation emitted is dependant only on the amount of the radioactive element present. The behaviour of radioactive materials is not affected by physical or chemical processes, e.g. temperature or atmospheric conditions.

Improve your grade

Radioactive elements

Foundation: Radioactive materials emit ionising radiation. Explain what is meant by 'ionising radiation'.

AO1 [3 marks]

Types of radiation and hazards

Types of ionising radiation

- There are three types of ionising radiation emitted by radioactive materials.

	Charge	Mass			
Alpha (α)	Positive (2⁺)	Heavy	Deflected by magnetic and electric **fields**	Highly ionising	Low penetration
Beta (β)	Negative (1⁻)	Very light	Deflected by magnetic and electric fields (in opposite direction to alpha)	Not as ionising as alpha	Medium penetration
Gamma (γ)	Neutral	No mass	Not deflected by magnetic or electric fields.	Low ionising power	High penetration

- As alpha, beta and gamma radiation go through air they ionise air molecules and lose energy.

- Alpha particles are massive and can easily knock off electrons, so lose energy quicker and travel less far (few cm in air).

- Beta particles have a range in air of about a metre, but can be stopped by about 3 mm of aluminium.

- Gamma rays are not stopped by air, but just spread out and become less intense. Thick lead is used to absorb gamma rays.

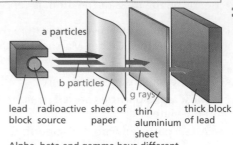

Alpha, beta and gamma have different penetrating properties.

- An alpha particle is a helium nucleus. It consists of 2 protons and 2 neutrons, so it has an atomic mass of 4 and a charge of +2. It is written as ^4_2He.

Higher

- A beta particle is a fast-moving electron, so it has an atomic mass of zero and a charge of −1. It comes from the nucleus when a neutron changes into a proton and an electron. It is written as ^0_1e.

- Gamma rays are very-high-frequency electromagnetic waves.

Hazards of ionising radiation

- Ionising radiation can damage living cells. The damage depends on the type and intensity of the radiation.

- Ionising radiation collides with living cells and knocks elections out of the atoms, leaving positive ions.

- High intensity radiation may kill the living cells and cause tissue damage, and could lead to radiation sickness. Some high intensity radiation causes cells to become **sterile**.

- Lower intensity radiation can affect the cell's genetic makeup, causing **mutations** that could lead to cancer.

- Alpha particles are large and highly ionising, but do not pass through the skin. Inside the body alpha particles would be highly damaging, but are relatively safe outside the body.

- Beta and gamma are much more penetrating and will pass through skin, so are more dangerous outside the body despite being poorer ionisers.

- The unit of radiation absorbed **equivalent dose** is the **Sievert (Sv)**. One Sievert of alpha, beta or gamma produces the same biological effect, and is a measure of the possible harm done to your body.

- Oxygen, hydrogen, nitrogen and carbon are highly susceptible to ionisation and are abundant in the body.

- Ions can interfere with the structure of DNA, causing it to behave incorrectly and damage living cells.

Higher

- Once atoms are ionised by radiation and form ions, these charged ions can break and make chemical bonds, therefore changing molecular structure.

Ideas about science

You should be able to:
- suggest ways of reducing the risks from ionising radiation
- interpret and discuss information on the size of risks, presented in different ways.

Improve your grade

Hazards of ionising radiation

Higher: Explain why it is more dangerous to inhale (breathe in) an alpha emitter than a beta emitter.

AO1, AO2 [4 marks]

Radioactive decay and half-life

Radioactive decay

- When a radioactive nucleus emits an alpha or beta particle, the number of protons and neutrons changes and it becomes a new element.

- The radioactivity of a material decreases over time because the amount of radioactive nuclei decreases. This is called **radioactive decay**.

- The **activity** of a radioactive sample only depends on the number of unstable nuclei present.

- Many large nuclei are unstable and emit alpha particles. For example, uranium-238 decays to the element thorium with the emission of an alpha particle.

- Carbon-15 is an unstable form of carbon that emits a beta particle and becomes nitrogen.

- Emitting gamma rays does not change one element to another.

- The new element is known as the **daughter product**, and may or may not be radioactive.

- Many radioactive elements belong to a **decay chain**, in which the first radioactive element decays into a second element, which then decays into a third element, and so on.

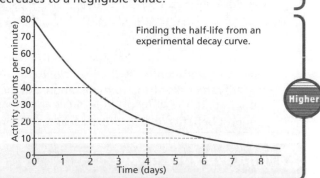

Alpha decay.

Carbon decay.

- An unstable nucleus is in an *energetic state* – it has too much energy. It needs to lose energy to become more stable.

- When nuclei emit alpha or beta particles they change into more stable nuclei, but may still be in an energetic state. So they often emit gamma rays as well, to reduce their energy.

- Scientists represent nuclei in the form $^M_A X$, where X is the chemical symbol for the element, M is the mass number (number of protons + number of electrons) and A is the atomic number (number of protons).

- An equation for alpha decay is: $^{238}_{92}U$ → $^4_2He + ^{234}_{90}Th$ 238 – 4 / 92 – 2

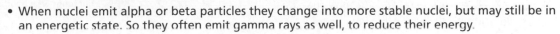

- An equation for beta decay is: $^{15}_6C$ → $^0_{-1}e + ^{15}_7N$ 15 – 0 / 6 + 1

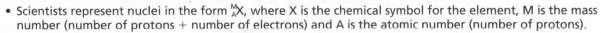

- The total mass and the total atomic number (or charge) must be the same on both sides of the equation.

Half-life

- The **half-life** of a radioactive element is the time taken for half the nuclei in a sample to decay. The half-life is specific to each radioactive element.

- Half-lives can vary from fractions of a second to millions of years.

- The activity of a radioactive source (the amount of radiation emitted) is a measure of its rate of decay.

- When there are plenty of radioactive nuclei present at the beginning, the rate of decay is faster than when most of the nuclei have already decayed.

- The activity can never reach zero – it just continuously decreases to a negligible value.

- Scientists can find the half-life of a sample by recording the radioactive count rate over time and plotting a graph. They read off the time it takes the count rate to drop from 80 to 40, 40 to 20, 20 to 10, etc., and calculate the average.

- So, if you started with a count rate of 120 counts per minute (cpm), after three half-lives the count rate would be: $\frac{1}{8} \times 120 = 15$ cpm

Finding the half-life from an experimental decay curve.

Improve your grade

Half-life

Higher: Radon-220 decays by alpha emission with a half-life of 52 seconds. The initial activity is 640 counts per second. How long will it take for the activity to become 80 counts per second?

AO1, AO2 [3 marks]

Uses of ionising radiation and safety

Uses of ionising radiation

1 *Treating cancer* – ionising radiation can kill cells, so can be used to kill cancerous cells. This is known as **radiotherapy**. Gamma radiation is usually used. Some healthy tissue around the tumour can be damaged, so the radiation must be focused on the tumour.

2 *Sterilising medical instruments* – these can be irradiated with gamma radiation to kill bacteria.

3 *Sterilising food* – as soon as fresh food is picked and ready to transport, micro-organisms will start the decay process. If the food is irradiated with gamma radiation the micro-organisms will be killed. This makes the shelf life of the food much longer.

- For sterilisation, the radiation must penetrate the packaging and be capable of killing bacteria – so gamma emitters are used.

4 *Detecting tumours*: brain and other tumours can be detected using a **radioactive tracer**. A gamma emitter with a half-life of a few hours is injected; the radiation is detected from outside to build up a computer image of the tumour.

EXAM TIP
Always explain which type of radiation you need and whether you need a long or short half-life for all uses of radioactivity.

- Radioactive tracers are usually beta or gamma emitters, as they must be able to penetrate skin and tissue. The half-life needs to be a few hours, so that it has time to reach the affected parts of the body in sufficient quantities, but not last so long that it damages the body.

Keeping people safe

- Exposure to radiation is called **irradiation**.

- People are exposed to radiation all the time and the risk to health is usually insignificant. However, it does depend on the level of radiation and the length of exposure.

- The dose equivalent in Sieverts can be used to evaluate the level of risk from radiation and what harm may be done.

Remember!
A **hazard** is anything which may cause harm. The **risk** is the chance, high or low, that somebody could be harmed by the hazard.

From the sky
About 400 000 cosmic rays pass through us each hour

From the air
30 000 atoms of radioactive gases breathed in disintegrate in our lungs each hour

From food
15 million potassium-40 atoms disintegrate inside our bodies each hour

From soil and building materials
More than 200 million gamma rays pass through us each hour

We are exposed to ionising radiation all the time.

- **Contamination** means that a surface or person is in contact with radioactive material.

- Radioactive waste can be contained to prevent contamination. If radioactive waste cannot be contained it must be diluted to safe concentrations.

- High-level contamination, such as fallout from a nuclear explosion, will need more intervention, e.g. administering iodine to affected people.

- People who work with radioactive sources, e.g. radiographers and workers in nuclear power stations, regularly need to have their level of exposure monitored.

- A film badge can monitor radiation. The level of exposure is measured by assessing how black the film has become.

- Lead shielding in the form of aprons or protective walls is used to protect radiographers.

Ideas about science

You should be able to:

- identify examples of risks which arise from a new scientific or technological advance, such as using gamma radiation to sterilise food and medical instruments
- identify the risks of the contamination and the benefits of using radioactive materials in a given situation, to the different individuals and groups involved
- suggest benefits of activities, such as radiotherapy, that are known to have risk.

Improve your grade

Uses of ionising radiation

Higher: Radioactive tracers can be used in many different applications. Which of the following radioactive isotopes would be most suitable for studying plant nutrition?

Explain your choice.

Isotope	Type of radiation	Half-life
Phosphorus-32	Beta decay	14 days
Nitrogen-16	Beta decay	7 seconds
Bismuth-210	Alpha decay	5 days

AO1 [3 marks]

Nuclear power

Energy from the nucleus

- **Nuclear fission** releases energy by a heavy nucleus, e.g. uranium, splitting into two lighter nuclei.

- **Nuclear fusion** releases energy by two light nuclei, e.g. hydrogen, combining to create a larger nucleus.

- Materials that can provide energy by changes in the nucleus are known as **nuclear fuels**.

- Nuclear fission and fusion release much more energy than chemical reactions involving a similar mass of material. This is because the energy that holds nuclei together (**binding energy**) is much larger than the energy that holds electrons in place.

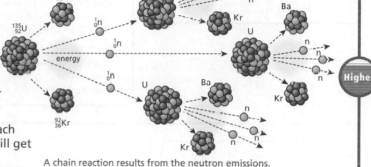

- In nuclear fission, a neutron is fired at a uranium or plutonium nucleus to make it unstable. The nucleus breaks down into two smaller nuclei of similar size, and releases some more neutrons.

- The neutrons released in the fission reaction can go on to initiate more fission reactions. This is known as a **chain reaction**.

- Only one neutron from each fission needs to go on to initiate the next fission.

- It is necessary to have a critical mass of nuclear fuel for the chain reaction to be viable.

- More and more neutrons will be released in each subsequent reaction, and the chain reaction will get out of control unless the number of neutrons is controlled.

A chain reaction results from the neutron emissions.

- Energy released in nuclear fission and fusion is calculated using $E = mc^2$, where E is the energy in joules, m is the mass in kg and c is a constant equal to the speed of light in a vacuum or 3×10^8 m/s.

Nuclear power generation

- About a sixth of the UK's electricity is generated in nuclear power stations. They all use nuclear fission.

- Nuclear fission produces radioactive waste, which must be disposed of carefully.

- Nuclear wastes are categorised according to their level of risk:

  - **Low level waste**, e.g. contaminated paper, clothing, is not dangerous to handle but should be disposed of with care. It is burnt and sealed in containers before being buried in landfill.

  - **Intermediate level waste**, e.g. chemical sludges, reactor parts, is more radioactive and needs shielding. Waste with a longer half-life is buried deep underground.

  - **High level waste**, e.g. spent fuel rods, is highly radioactive. Some of this waste is mixed with molten glass and contained in stainless steel drums before careful storage.

- In a nuclear reactor, the **fuel rods** contain pellets of uranium. Neutrons cause the fuel to undergo fission. The energy is released as kinetic energy of particles (heat).

- A **coolant** (gas or liquid) circulated around the reactor absorbs the heat and transfers it to a steam generator. Electricity is then generated in the same way as a conventional power station.

- **Control rods** (usually made of boron) absorb some of the neutrons. The control rods can be raised or lowered to control the fission rate.

Harnessing fusion energy

- Fusion produces a lot more energy per kg than fossil fuels. Its by-products are not radioactive and it does not release carbon dioxide into the atmosphere.

- Isotopes of hydrogen are readily available and only small amounts are needed.

- Despite its great potential, more energy is consumed producing fusion reactions than is released by it.

Improve your grade

Energy from the nucleus

Foundation: Explain the difference between nuclear fission and nuclear fusion.   *AO1, AO2* [4 marks]

P6 Summary

Atoms have neutrons and protons in the nucleus, which are held together by a strong nuclear force. The nucleus is surrounded by electrons.

The Rutherford-Geiger-Marsden scattering experiment gave evidence for the structure of the atom.

An isotope is an atom of an element with the same number of protons but a different number of neutrons in the nucleus.

Ionisation occurs when radiation collides with an atom and knocks electrons out of orbit, leaving a positively charged ion.

Nuclear radiation

When an unstable nucleus emits alpha or beta particles it decays to become more stable – it becomes a new element.

An equation for alpha decay is: $^{238}_{92}U \rightarrow {}^{4}_{2}He + {}^{234}_{90}Th$

An equation for beta decay is: $^{14}_{6}C \rightarrow {}^{0}_{-1}e + {}^{14}_{7}N$

The activity of a radioactive source decays over time.

The half-life is the time taken for the activity of a radioactive sample to halve, or the time taken for the number of radioactive nuclei to halve.

Unstable nuclei are radioactive and emit ionising radiation.

Background radiation comes from natural and man-made sources.

Radioactivity is completely random and is not affected by chemical or physical changes.

There are three types of ionising radiation:

- Alpha – helium nucleus, massive, 2+ charge, highly ionising, low penetration. $^{4}_{2}He$
- Beta – fast-moving electron, low mass, 1+ charge, medium ionising power, medium penetration. $^{0}_{-1}e$
- Gamma – high frequency electromagnetic waves, low ionising power, high penetration.

Ionisation can cause damage to living cells – they could be killed or become cancerous.

Ions (formed by radiation colliding with atoms) can react with other chemicals.

Uses and safety

Exposure to ionising radiation is called irradiation.

Contamination occurs when you are in contact with radioactive materials.

Background radiation causes irradiation and contamination all the time.

Radiation dose (measured in Sieverts) is a measure of the possible harm done to our bodies. The higher the radiation dose, the greater the risk.

People who work with radioactive materials must monitor their exposure carefully.

Ionising radiation is used to treat cancer, sterilise medical equipment and food, and as a tracer in the body.

Choosing the radioactive source involves thinking about both the type of radiation emitted and the half-life.

The longer the half-life, the longer the source will be considered dangerous.

Nuclear fuels can release a lot more energy than chemical fuels.

Some of the binding energy is released as there is a change of mass.

Energy released $E = mc^2$.

Nuclear power

Nuclear power stations produce radioactive waste.

Low level waste, intermediate level waste and high level waste are disposed of in different ways.

Nuclear fusion occurs when two light nuclei, e.g. hydrogen, come close enough together to fuse and form a heavier product, e.g. helium. This releases energy.

Nuclear fission occurs when a large unstable nucleus, e.g. uranium or plutonium, splits into two similar-sized smaller nuclei. This gives off large amounts of energy and some neutrons.

The neutrons go on to initiate further fission of more nuclei in a chain reaction.

In nuclear power stations, the number of neutrons which go on to cause more fission in the fuel rods is managed by control rods.

The fuel rods are surrounded by a coolant which transfers heat energy to water to produce steam. The steam turns a turbine to generate electricity.

The solar day

The rotation of the Earth

- The Earth orbits the Sun.

- The Earth spins on its axis so that the Sun appears to move from the east to the west across the sky.

- The time that passes between the Sun appearing at its highest point on one day and the next is a **solar day**.

- A solar day is 24 hours long – if you looked at where the Sun is in the sky and waited 24 hours, it would be in the same position.

- The Moon and stars also appear to travel from east to west across the sky.

- The time it takes for the Moon to appear in exactly the same position between one day and the next is more than 24 hours.

  - Twenty-four hours for the Earth to rotate once, plus around 50 minutes for the Earth's rotation to catch up with where the Moon has moved to in its orbit around the Earth.

- The time taken for stars to appear to travel east-west across the sky is very slightly less than 24 hours.

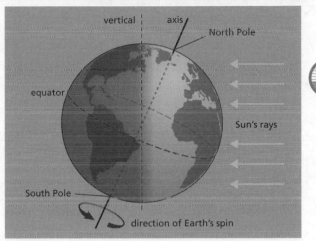

The Earth spins on its axis so that the Sun appears to rise in the east and set in the west.

Remember!
The Sun and stars are **NOT** travelling around the Earth, they just look like they do because the Earth is rotating on its axis.

Sidereal days

- A **sidereal day** is the time it takes for the Earth to undergo a complete rotation of 360 degrees. It is 23 hours and 56 minutes long and can be seen as the time it takes for stars to appear in the same position from one night to the next.

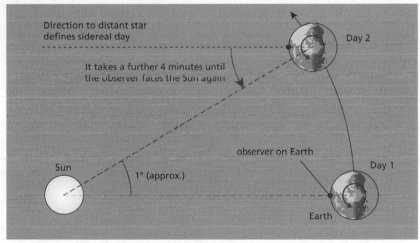

After one full rotation of the Earth, the Sun is not quite overhead again.

- The solar day is longer than the sidereal day because not only is the Earth rotating on its axis but it is also orbiting the Sun.

- In the time it takes the Earth to complete one full rotation, it has moved a bit through space so needs to rotate a bit further on its axis for the Sun to appear in the same position as it did the day before. This extra rotation takes 4 minutes, making the solar day 24 hours long.

EXAM TIP

When wording your answers make sure you talk about the Earth rotating – **not** the Sun and stars moving around the Earth. It is fine to say the Sun and stars **appear** to move.

Improve your grade

Sidereal days

Higher: Explain the difference between the length of a solar day and the length of a sidereal day.

AO1 [4 marks]

The Moon

The phases of the Moon

- The Moon orbits the Earth at a distance of around 384 000 km.

- The Moon does not give off its own light like the Sun but can be seen because the light from the Sun reflects off it.

- Only one side of the Moon is lit by the Sun at any time and, because the Moon orbits the Earth, how much of this lit side we can see changes.

- The Moon appears to change shape due to the change in the amount of the lit side we can see. These shapes are called the phases of the Moon.

- When the Moon is almost directly between the Sun and the Earth, the lit face of the Moon points away from the Earth and the Moon is barely visible; this is called a new Moon.

- When the Earth is almost directly between the Sun and the Moon an observer on Earth can see the full, lit face of the Moon; this is called a full Moon.

- The time between one full Moon and the next is 29.5 days; this is called a **lunar month**.

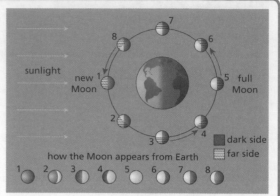

One side of the Moon is lit by the Sun. As the Moon orbits the Earth, we see the 'phases' of the Moon.

Remember!
A new Moon is **NOT** caused by the Moon being in the Earth's shadow but is because we can only see the dark side of the Moon.

Eclipses

- When the Earth, Sun and Moon are directly in line it creates an eclipse.

- A **solar eclipse** occurs when the orbit of the Moon takes it directly between the Earth and the Sun and the Moon's shadow falls on the Earth. Each solar eclipse can only be seen at certain places because the Moon is smaller than the Earth so its shadow covers a relatively small area.

- A **lunar eclipse** is when the Moon's orbit causes it to pass into the Earth's shadow. A lunar eclipse covers the entire Moon and can be seen from anywhere on Earth where the Moon would be visible.

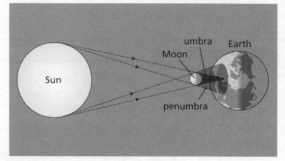

Observers in the area of a total shadow, called the umbra, see a total eclipse. In an area of partial shadow, the penumbra, there is a partial eclipse.

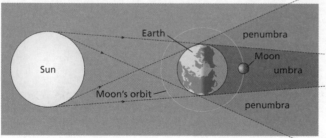

A lunar eclipse occurs when the Moon is in the Earth's shadow.

- Eclipses do not occur every full Moon and new Moon because the Moon does not orbit the Earth in the same plane as the Earth orbits the Sun.

- For an eclipse to occur the Earth, Sun and Moon must line up perfectly, which only occurs on average two or three times a year.

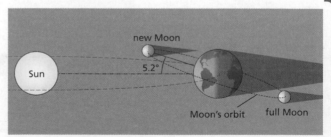

The Moon's orbit is inclined at an angle to the Earth's.

Improve your grade

The phases of the Moon

Foundation: During a new Moon the Moon is not visible in the night sky. What causes this?

AO1 [2 marks]

The problem of the planets

The naked eye planets

- Mercury, Venus and Mars can all be seen from Earth because they are the closest three planets to Earth so the light reflected from them can be seen with the naked eye.

- Jupiter and Saturn, although further away, can also be seen with the naked eye because they are very big planets.

Wandering stars

- The word 'planet' comes from the Greek word 'wandering' and was used because although generally moving with the stars, the planets appear to wander across the sky.

- As planets orbit the Sun their distance from the Earth changes and so their apparent brightness changes; when closer to Earth they appear brighter, when further away, dimmer.

- As the Earth rotates on its axis all the planets appear to move east-west with the stars.

- But the planets are closer than the stars and also orbit the Sun so their position compared to the background of stars also changes, and they occasionally appear to move in the opposite direction.

- When planets appear to move backwards, we call this **retrograde motion**.

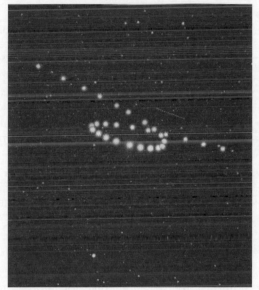

The path of Mars shown by superimposing 35 pictures taken at the same time of night, one week apart.

- The planets orbit the Sun at different speeds.

- Those closer to the Sun move faster in their orbits than those further away.

- This difference in the speed of different planets and their position relative to the Earth and the Sun means that sometimes planets appear to be moving towards us and sometimes they appear to be moving away.

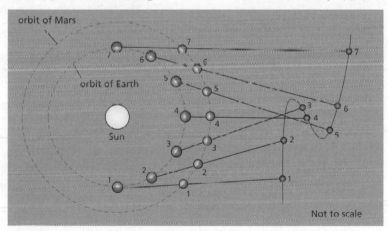

As the Earth moves faster than Mars, Mars appears to go backwards against the background of stars.

- When Mars is on the opposite side of the Sun to the Earth it will appear to be moving faster as the two planets are moving in opposite directions, and it will travel a further distance compared to the background of fixed stars.

- As the Earth approaches Mars it will catch up with it and then overtake it, which means that for a period of time Mars will appear to be going backwards (retrograde motion). This can be seen in the photo above and in the diagram on the right.

> **Remember!**
> The planets do not really change their speed or the direction of their orbit – they just appear to do so when compared to the background of 'fixed' stars, due to the relative motion of themselves and an observer on Earth.

Improve your grade

Planets

Foundation: Although smaller than Jupiter, the planet Venus appears brighter. Suggest why.

AO2 [2 marks]

Navigating the sky

Finding stars

- The Pole Star (Polaris) can be found above the North Pole and can be seen all year round.

- The stars visible in the night sky change depending on the time of the year, due to the position of the Earth in its orbit around the Sun.

- At night we are facing away from the Sun.

- As the Earth orbits the Sun, the night sky looks out onto a different part of the Universe each night. Across a full year the night sky changes a little bit each night so across the year it looks out to the Universe in all directions.

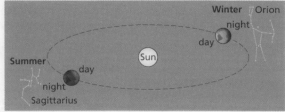

We view the night sky in summer in the opposite direction to the night sky in winter.

- E.g. during summer we are on one side of the Sun and during winter we are on the other, so the night sky shows the other side of the Universe and different constellations are visible.

- The position of stars and other astronomical objects are described with two angles, the angle of **declination** and the angle of right ascension.

- The angle of declination tells you how high above the equator to look; a star directly above the equator would have a declination of zero degrees and a star directly above the North Pole would have a declination of 90 degrees. Stars that are south of the equator have a negative declination.

- The angle of **right ascension** tells you which way to look east-west.

- Together the two angles give you coordinates to look at to find the star.

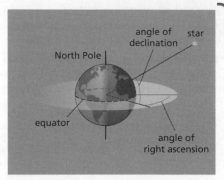

Locating a star by angular coordinates.

The celestial sphere

Higher

- The **celestial sphere** is an imaginary dome that stretches over the Earth and has the light from every astronomical object projected onto it.

- It has a north pole above the North Pole and an equator above the equator.

- The sphere is used as the reference point for the coordinate system.

- The angle of declination is the angle made between the object and celestial equator directly below, or above, the object.

- The angle of right ascension is measured from the point where the Sun moves to the Northern Hemisphere – this can be seen in the diagram as the point where the **ecliptic** (the plane of the Earth's orbit and path of the Sun over the year as viewed from Earth) crosses the equator and takes place at the spring (vernal) equinox.

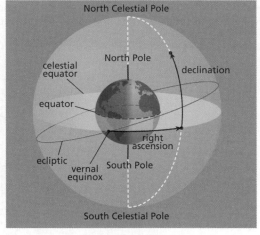

The celestial sphere.

- The angle of declination is measured in degrees, minutes and seconds.

- The angle of right ascension is measured in hours, minutes and seconds. This is because the Earth and its celestial sphere take 24 hours to rotate 360 degrees so 1 hour is equivalent to 15 degrees.

Improve your grade

Using angles of ascension and declination

Foundation: Compare the positions of the two stars described in the table below.

	Angle of right ascension	Angle of declination
Star A	30 degrees	−10 degrees
Star B	30 degrees	+75 degrees

AO2, AO3 [3 marks]

P7 Further physics – studying the Universe   **Student book pages 134–135**

Refraction of light

The speed of light

- Light travels as a wave.

- The speed of a wave is affected by the medium it is travelling through.

- When a wave passes from one material to another its speed will change.

- In a vacuum light travels at 300 000 000 m/s but it slows down slightly when it enters the atmosphere and slows down even further when travelling through glass.

- The diagram shows water waves travelling from deep water into shallower water. When this happens the waves slow down.

- The number of waves leaving the deep water per second must be the same as the number entering the shallow water so the frequency stays the same.

- Because the frequency cannot change, the change in speed must cause a change in wavelength – when the speed gets slower, the wavelength gets shorter.

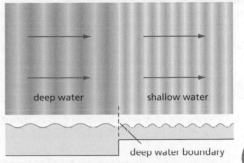

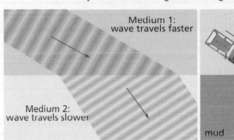

Water waves travel more slowly in shallow water, so they slow down and get closer together.

- When waves hit a boundary between two materials at an angle they change direction as they pass from one medium into another. This is called **refraction**.

- This can be visualised as being similar to if a car drives from grass into mud; the wheel that enters the mud first will slow down and this will cause the car to change direction.

- For light we show this with a **ray diagram**. As light enters the glass block it slows down and refraction causes it to deviate towards the **normal**. When it leaves the block, it bends away from the normal. The normal is a construction line at right angles to the surface of the glass.

- It is this refraction that is used by a lens to bring light into focus.

The part of the wave front that hits the boundary first will change speed first and this effect will change the direction of the wave.

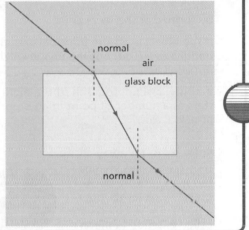

Refraction of light as it passes through glass.

Lenses

- By shaping the glass block we can make light change direction to the direction we require.

- A **convex lens** is shaped in such a way so that it is thicker in the middle than at the edges.

- Light entering the lens will be turned inwards as it deviates towards the normal.

- On leaving the lens the light deviates away from the normal but the shape of the lens means that this also turns the light inwards.

- In this way the light rays all **converge** on a single point, called the **focal point**, and this is why this type of lens is also called a converging lens.

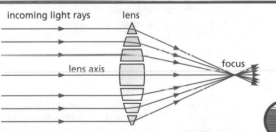

The light ray moves towards the normal as it hits the glass and away from the normal as it leaves. So shaping the glass to form a convex lens makes the light converge to a focus.

EXAM TIP

You will need to be able to draw diagrams showing how a converging lens brings light to a focus. When doing so make sure you use a ruler and that the direction changes of your light rays are sudden, not curves.

⦿ Improve your grade

Refraction

Foundation: Light travels slower in water than in air; use this information to complete the diagram by adding ray lines to show how the observer is able to see the bottom of the cup.

AO2 [3 marks]

Forming an image

The power of a lens

- **Convex lenses** bring rays of light together. Parallel rays entering the lens are brought together at the **focal point**.

- How much the rays are turned inwards depends on the material of the lens and how curved it is.

- The more curved the surface of the lens, the further the rays are tuned inwards and the shorter the **focal length**.

- The **power** of a lens is measured in **dioptres**. The more powerful a lens, the shorter the focal length.

$$\text{Power (dioptres)} = \frac{1}{\text{focal length (metres)}}$$

- So a more curved lens is more powerful and has a shorter focal length.

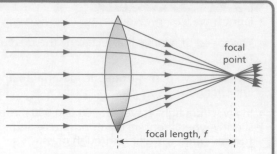

A more powerful lens has a shorter focal length.

Focusing light from astronomical objects

- Light rays entering the eye from close up objects are not parallel but astronomical objects are so distant that the rays of light from them reaching Earth are effectively parallel sets of rays.

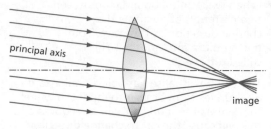

The light rays from a nearby object, A, form a large angle at the eye. The further away the object, B, the smaller the angle. A star is so far away that the light rays from the star are parallel as they reach the eye.

- When a lens is used to focus the light from a distant star which is on the **principal axis** of the lens, the image will be formed at the focal point.

- If the object is off the principal axis, the rays will still be bought to a focus but the image will be formed off the principal axis.

- In this way a photograph of the night sky will show the stars as a number of point images positioned around the principal axis.

The light rays from a star are parallel. The lens brings the rays to a focus, forming a point image of the star.

- Stars are so far away they form point images but many astronomical objects look larger than a point. These objects are called **extended objects**.

- Galaxies, the Sun, and planets and moons in our solar system are seen as extended objects.

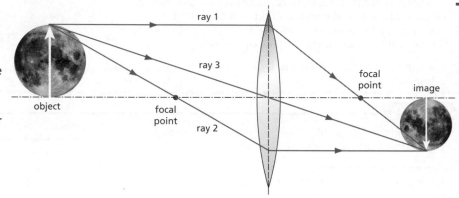

- A ray diagram can be used to show how an image of an extended object is formed. Draw three rays leaving the top of the object.

Rays of light from a point on the extended object (such as the Moon) are brought together again by the lens to form a focused image. This is not drawn to scale: the Moon is much further away and its image will be much smaller.

- Ray 1 hits the lens parallel to the principal axis and is refracted through the focal point on the image side of the lens.

- Ray 2 passes through the focal point on the object side of the lens, hits the lens and is refracted parallel to the principal axis.

- Ray 3 passes through the centre of the lens and continues in a straight line.

- The image of the top of the object is formed where these three lines intersect.

- The image of the bottom of the object sits on the principal axis.

EXAM TIP

You will need to be able to draw all of the diagrams on this page. When doing so remember to add an arrow to each ray to show the direction it is travelling. Only draw one arrow per straight line section.

Improve your grade

The power of a lens

Foundation: A magnifying glass has a focal length of 10 cm. Calculate the power of the lens. *AO2* [3 marks]

The telescope

The simple refracting telescope

- A simple telescope contains two converging (convex) lenses, an **eyepiece** lens and an **objective lens**.

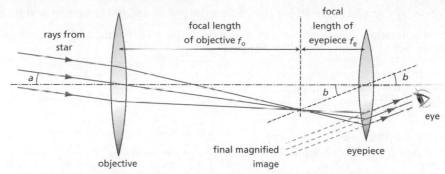

rays from
star

a

focal length
of objective f_o

focal
length of
eyepiece f_e

b

b

eye

final magnified
image

eyepiece

objective

The arrangement of the objective lens and eyepiece lens in an astronomical telescope.

- The objective lens is a low power lens with a long focal length. It has a large diameter to collect more light and make a brighter image.

- The eyepiece lens is smaller but with a high power and short focal length. It acts like a magnifying glass to magnify the image formed by the objective lens.

Magnification

- The function of a telescope is to make objects appear larger; it works by making the angle between the light rays and the axis of the telescope larger.

- In the diagram angle b is larger than angle a, and the ratio of these angles gives the angular magnification of the telescope. Magnification $= \dfrac{b}{a}$

- In this diagram angle b is 2.5 times the size of angle a. This makes stars look 2.5 times as far apart and extended objects appear 2.5 times the size than when looked at with the naked eye. So, this telescope has a magnification of 2.5.

- The magnification of a telescope depends on the ratio of the power and therefore the focal lengths of the lenses.

- The telescope in the diagram has a magnification of 2.5, the focal length of the objective lens is 2.5 times that of the eyepiece lens.

- Magnification $= \dfrac{\text{focal length of objective lens}}{\text{focal length of eyepiece lens}}$

- Example: A telescope is made from two lenses; the eyepiece lens has a power of 10D and the objective lens a power of 2D. What is the magnification?

First use the power to find the focal length (see page 44) $f = \dfrac{1}{\text{power}}$

Focal length of objective $= \dfrac{1}{2} = 0.5$

Focal length of eyepiece $= \dfrac{1}{10} = 0.1$

Magnification $= \dfrac{\text{focal length objective}}{\text{focal length eyepiece}} = \dfrac{0.5}{0.1} = 5$

So the telescope is 5x magnification. The image will appear five times larger than if viewed with the naked eye.

- It can be seen from this example that the magnification could also be found directly from the powers by dividing the eyepiece power by the objective power, $\dfrac{10}{2} = 5$.

Improve your grade

Magnification

Higher: A refracting telescope has an eyepiece lens with a focal length of 5 cm and an objective lens with a magnification of 1 m. Find its magnification and explain what this means. *AO2* [4 marks]

The reflecting telescope

Why use mirrors instead of lenses?

- Most astronomical telescopes use concave mirrors instead of convex objective lenses.
- A concave mirror is lighter and easier to support than a convex lens of the same size and focal length.
- A mirror can be made larger than a lens and so is able to capture more light and can see fainter objects. This allows it to detect light from objects further away.

- The concave mirror in a reflecting telescope does the same job as the objective lens does in a refracting telescope – it brings parallel rays of light to a focus.
- Light reflecting off a surface obeys the law of reflection; the angle of reflection is equal to the angle of incidence.
- Because the mirror is curved, parallel rays hitting the surface do not hit the mirror along the normal so are not reflected back the way they came; the rays are reflected inwards towards a single focal point.
- In the reflecting telescope a secondary mirror is used to reflect the light towards an eyepiece lens, which performs the final magnification of the image in the same way as the one in a refracting telescope.

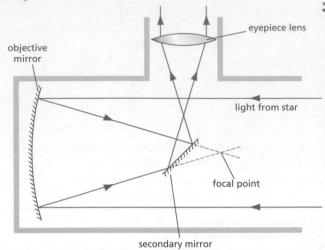

A Newtonian reflecting telescope produces an image at infinity.

- The magnification of a reflecting telescope is found in the same way as a refracting telescope (focal length of objective / focal length of eyepiece).

Chromatic aberration

- Concave mirrors have a number of advantages over convex lenses.
  - It is easier to manufacture a large mirror than a large lens.
  - A large mirror can be much thinner than a large lens so is lighter.
  - A large mirror can be supported all along its rear side but a lens

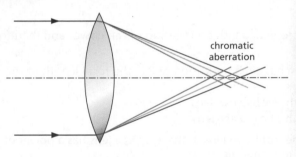

A lens has a different focal length for every colour of light, so a simple refracting telescope produces images with coloured fringes.

can only be supported at the edge, otherwise the light passing through the lens is blocked.

- The above advantages make the use of mirrors for the construction of large telescopes the obvious choice due to relative ease and reduced cost. The other key advantage of mirrors is that they do not cause **chromatic aberration** like a lens does.

- When light passes through a lens it is refracted towards the normal, however different colours of light are refracted by different amounts and it is this effect that causes the spectrum when white light passes through a prism (see page 48).

- For telescopes this means that the light is separated into different colours and the image will appear coloured where it should not be, e.g. an image of the Moon appears coloured around the edges.

- Isaac Newton solved this problem by using a mirror in place of the objective lens when he built the Newtonian reflecting telescope in 1668.

> **Remember!**
> A **refracting** telescope uses a **convex** lens for the objective and a **reflecting** telescope uses a **concave** mirror.

Improve your grade

Why use mirrors?

Foundation: Reflecting telescopes are not affected by chromatic aberration. Explain another reason why most astronomical telescopes are reflecting telescopes instead of refracting telescopes.

AO1, AO2 [3 marks]

Diffraction

What factors affect how much a wave is diffracted?

- Telescopes need to be large so that they can detect weak signals but they also need to be large so that they can produce detailed images. This is because of diffraction. When waves pass through a gap or round an obstacle they spread out; this is called **diffraction**.

- Because the wavelength of light is very short we don't normally notice the diffraction of light but if light shines through a very small gap it will be diffracted. We notice this most when the gap is around the same width as the wavelength of light.

Water wave diffraction

- The amount of diffraction for any wave depends on its wavelength and the diameter of the gap.

- A narrow gap will cause more diffraction.

- The longer the wavelength, the greater the amount of diffraction. Radio waves have a long wavelength and can be diffracted by hills into valleys; this is why you can receive

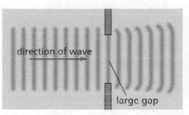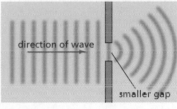

Diffraction is greater when the gap size approaches the wavelength.

a radio signal even if you don't have direct line of sight to the transmitter. You can't see the transmitter because light has a much shorter wavelength and is not diffracted by the hills.

- Not all telescopes detect light. Many telescopes detect other types of radiation, for example a radio telescope detects long wavelength radio waves and an X-ray telescope detects wavelengths much shorter than light.

Diffraction and images

- When radiation enters a telescope it is diffracted by the aperture of the telescope. The **aperture** of a telescope is the size of the objective lens or objective mirror.

- Because the aperture is circular, a diffracted image of a star will appear blurry with a bright central disc surrounded by circular rings instead of a sharp point.

- To produce sharp images the aperture must be very much larger than the wavelength, so the longer the wavelength of the radiation being detected, the larger the aperture needs to be.

- As can be seen in the star photographs (right), a larger aperture (B and C) produces greater **resolution** and sharper images.

- The problem of diffraction is particularly noticeable with radio telescopes.

- Radio waves have a long wavelength so radio telescopes are very large and are often joined together in an array to create a telescope with an aperture hundreds of metres across.

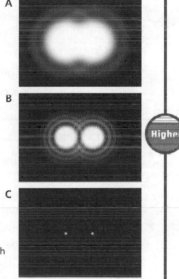

These simulated pictures of a double star were taken with a large telescope, using apertures of different diameters. **A** uses the smallest aperture and **C** the largest aperture.

Improve your grade

Diffraction and telescopes

Higher: The table below gives some information about telescopes. Use this information to put the telescopes in order of how much they are affected by diffraction, from the most affected to the least affected.

Telescope	Radiation detected	Wavelength detected	Aperture size
James Clerk Maxwell	Microwave	1 m	15 m
Proposed ELT	Visible light	5×10^{-7} m	42 m
Hopkins	Ultraviolet	1×10^{-7} m	1 m
Arecibo	Radio waves	100 m	300 m

AO2, AO3 [3 marks]

Spectra

Forming a spectrum

- Stars and galaxies emit radiation at many wavelengths. A star's spectrum shows how much energy of each wavelength it emits.

- Studying the spectrum of radiation emitted by a star can give us information about its composition and temperature.

- The spectrum of visible light from the Sun can be seen in a rainbow.

- We can separate light into the spectrum of colours by using a prism. Different coloured light has different wavelengths and different wavelengths are refracted by different amounts.

- White light is a mixture of all the colours of the spectrum. As white light enters the prism it is refracted and changes direction.

- Different colours change direction different amounts, splitting white light into the colours of the visible spectrum: red, orange, yellow, green, blue, indigo, and violet.

A rainbow is a spectrum of the Sun's light, created by raindrops acting as prisms.

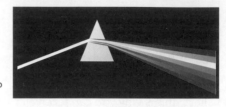

White light is separated into its component colours.

- The shorter the wavelength of light, the more it is refracted. This means that violet light is refracted the most and deviates the most from the original path.

- Longer wavelength red light is refracted the least and deviates least from the original path.

- A spectrum can also be produced by a **diffraction grating**.

- A diffraction grating is either a series of thousands of finely spaced gaps that light shines through or a series of finely spaced lines drawn on a reflective surface that light reflects off.

- A diffraction grating would typically have thousands of lines per cm.

- Different colours of light striking or passing through a diffraction grating are diffracted by different amounts.

- Different colours of light are visible from different angles; this affect can be seen when looking at the surface of a CD.

A CD acts as a reflective diffraction grating.

Quasars

- When astronomers discovered **quasars** they noticed that their spectrum was highly shifted to the red end of the spectrum.

- This very large **redshift** meant that the quasar must be moving away from us at a very high speed and, according to the theories of an expanding Universe, must therefore be very far away.

- Some astronomers didn't accept this theory, believing that we should not be able to see objects so far away so there must be another reason for the amount of redshift observed.

- It took 20 years and the discovery of many more quasars for it to become accepted that quasars are among the brightest, most powerful and most distant objects in the Universe.

Ideas about science

You should be able to:

- Explain that evidence can be interpreted in several different ways.

- Explain that a theory becomes accepted when predictions based on the theory are proven correct by sufficient evidence and when the scientific community have studied and agreed with the findings.

- Explain that even accepted theories can change if new evidence or a better explanation comes to light.

Improve your grade

Looking at evidence

Foundation: Just after Isaac Newton earned his maths degree at university he was playing with prisms at home and saw that a different coloured light came out to that which went in. He proposed that white light was a mixture of different colours. Suggest why Newton's ideas were not generally accepted straight away.

AO2 [3 marks]

The distance to the stars

Parallax

- It is difficult to work out how far away stars are – two stars that appear equally bright could be one bright star a long way away and a dimmer star that is closer.

- One method we use to work out how far away stars are is called **parallax**. This method uses the effect that as we move a nearby object appears to move quickly but a distant object appears to move slowly. You see this when looking out of the window of a moving car, or when alternately closing one eye then the other.

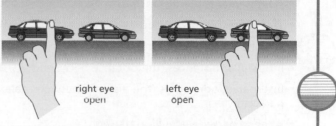

right eye   left eye
open   open

Look at a distant object through your right eye only. Point at it. Keep focused on the object but open your left eye instead. Your finger will appear to move.

- As the Earth moves around the Sun, nearby stars appear to move against the background of more distant stars.

- To find the distance to a nearby star two measurements are taken six months apart.

- We compare the position of the star against the background of very distant stars to find the parallax angle.

- The parallax angle is half the angle moved against the background of very distant stars in six months.

- Parallax angles are measured in seconds of an arc ("), (also referred to as arcseconds). 1 arcsecond = $\frac{1}{60}$ of a minute = $\frac{1}{60}$ of a degree. The smaller the parallax angle, the further away the star is.

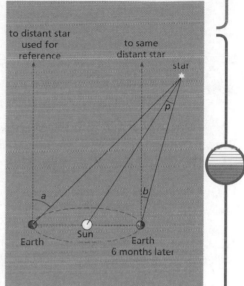

The reference star is so far away that its position in the sky doesn't change. The nearby star moves relative to the reference star, so angles a and b differ slightly (exaggerated here) and the parallax angle p, by geometry, is half of $(a - b)$.

The parsec and parallax

- The **parsec** is a unit of length defined using the parallax angle and is of a similar order of magnitude to the light year.

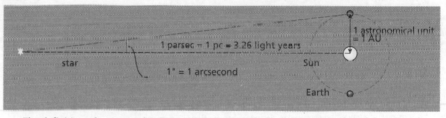

The definition of a parsec: the distance to a star is 1 parsec if the parallax angle is 1 second.

1 parsec = 3.26 light years.

1 light-year = distance travelled by light through space in a year.

$$\text{Distance (parsecs)} = \frac{1}{\text{parallax angle (seconds)}}$$

- Example: The closest star to us is Alpha Centauri, which has a parallax angle of 0.75". $\frac{1}{0.75} = 1.3$. Therefore Alpha Centauri is 1.3 pc away from Earth. This is about 300,000 AU so 300,000 times as far away as the Sun is from the Earth.

- Parallax angles to stars are very small; 1 arcsecond is approximately the angle you would have to move your eyes to look from one side of a penny to another if the penny was 1.5 km away.

- All stars have a parallax angle of less than 1 arcsecond. Compare this to the apparent sizes of the Moon and planets as seen from Earth:

  - A full moon has an angular diameter of 1800".

  - All the other planets in our solar system have an angular diameter of at least 2 arcseconds.

Remember!

Parallax can only be used to find the distance to nearby stars as the more distant the star is the less it appears to move so when the star is too far away its apparent movement is too small to be measurable.

Improve your grade

Parallax

Foundation: Explain why parallax can only be used to measure the distance to nearby stars. *AO2* [2 marks]

Distance and brightness in the Universe

Light years, parsecs and mega parsecs

- A **light-year** is the distance travelled by light in one year.
- The nearest star is just over 4 light-years (ly) away but the nearest galaxy is over 2 million light-years away.
- We also use parsecs to measure distance, with a parsec being equal to 3.26 ly.
- Interstellar distances (the distance between stars) are normally a few parsecs.
- Intergalactic distances (the distance between galaxies) are much greater and are measured in mega parsecs (Mpc).
- A mega parsec is a million parsecs.

Luminosity and apparent brightness

- The total power emitted by a star is its **luminosity**.
- This is the total energy emitted by the star in one second, across all wavelengths and in all directions.
- The luminosity of a star depends on:
  - Its size – a larger star has a greater surface area to emit radiation from.
  - Its temperature – the hotter the star the more radiation it emits. The temperature also effects the wavelength of the radiation it emits; hotter stars are bluer, cooler stars are redder.

- While luminosity refers to the amount of energy emitted by a star, its apparent brightness describes how much energy reaches Earth, i.e. how bright it appears to be.
- The more luminous a star the more light it emits but the further away a star the more the light from it has spread out by the time it reaches Earth, so the apparent brightness of a star depends on its luminosity and its distance from Earth.
- The star Sirius appears to be much brighter than the star Rigel, although Rigel has a much greater luminosity than Sirius. This is because Rigel is over 100 times further away.

In the constellation of *Orion* there are two exceptionally bright stars – the red supergiant *Betelgeuse* (top left) and the blue giant star *Rigel* (bottom right).

- Radiation spreading out from a point source obeys the inverse square law:
  - Doubling the distance from the source means the radiation is spread out over four times the area, so you receive $\frac{1}{4}$ of the radiation.
  - Tripling the distance spreads the radiation over nine times the area, so you receive $\frac{1}{9}$ of the radiation.
- If there are two stars of equal luminosity but one is 10 times further away it will appear to be $\frac{1}{100}$ as bright as the nearer star.

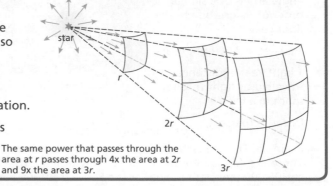

The same power that passes through the area at *r* passes through 4x the area at 2*r* and 9x the area at 3*r*.

Improve your grade

Luminosity and apparent brightness

Foundation: The table shows three well-known stars. Use the data in the table to order the stars according to which will appear brightest from the Earth, starting with the brightest.

Star	Luminosity	Distance from Earth
Alpha Centauri	1.5 L_o	4.3 ly
Sirius	25.4 L_o	8.6 ly
Rigel	126,000 L_o	860 ly

AO2, AO3 [2 marks]

Cepheid variables

Variable stars

- Some stars vary in brightness much more frequently and with a bigger variation than the Sun.

- The brightness of some of these stars can change rapidly, dimming and brightening again in just a few minutes, while some of them can take days or months to go through their cycle.

- The time taken to complete one full cycle is called the **period** of the star.

- The luminosity of a **Cepheid variable star** varies in a regular pattern, brightening and dimming again every few days.

- The top graph shows how the brightness of two different Cepheid variables changes over a period of 40 days; you can see how the luminosity of the brighter star does not change as fast as that of a fainter star.

- By studying a large number of Cepheid variable stars astronomers have been able to plot a graph (bottom) showing how the luminosity of a Cepheid variable star is connected to the period of the star.

- This relationship enables astronomers to estimate the distance to Cepheid variable stars.

- The study of Cepheid variable stars has been instrumental in determining the scale of the Universe.

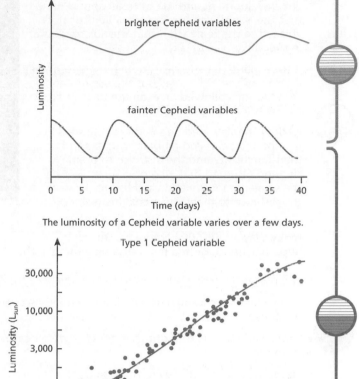

The luminosity of a Cepheid variable varies over a few days.

The period of a Cepheid variable is correlated with its luminosity.

Calculating distance

- Astronomers use parallax to calculate the distance to nearby Cepheid variable stars and then use this information and how bright the star appears to be to calculate the luminosity.

- They then link this to the period of the star to calculate how the luminosity depends on the period (see graph of Type 1 Cepheid variable).

- Astronomers are now able to estimate the distance to Cepheid variable stars that are too far away for the parallax method to be used; by studying the star to find its period they can work out its luminosity.

- Knowing the luminosity of a star, we know how much light it gives out; by comparing this to how much light actually reaches Earth we can calculate how far away the star is.

- When looking at Cepheid variable stars of the same period, the fainter the star, the further away it is. This is because they have the same period so they must have the same luminosity.

EXAM TIP

Higher – you will need to be able to explain how astronomers estimate the distance to a Cepheid variable star by looking at its period to find its luminosity and comparing this to its apparent brightness.

Improve your grade

Variable stars

Higher: An astronomer observes two Cepheid variable stars, one with a period of 12 days and one with a period of 20 days.

a Which star is most luminous?

b Explain how this can help the astronomer find the distance to the stars and what other information is needed.

AO2, AO3 [4 marks]

Galaxies

The night sky

- The development and use of telescopes revealed that the Milky Way consists of millions of stars. This led to the realisation that the Sun was a star in the Milky Way galaxy.

- It took many years for astronomers to determine the size of the Milky Way but the use of Cepheid variable stars allowed an estimate of the size of the galaxy to be determined.

- Harlow Shapley used Cepheid variable stars in groups of stars called **globular clusters** to find their luminosity and the distance to them. He then assumed that all globular clusters were the same luminosity and used this to measure the distance to many different clusters.

- Shapely came up with an estimated size for the Milky Way at 300 000 light years. Due to this huge size he suggested it was the only thing in the Universe.

A spiral galaxy like our own. The globular clusters shown are concentrations of thousands of old stars that surround the galaxy.

- The use of telescopes also led to the discovery of fuzzy bright clouds of light.

- These clouds are known as nebulae and their discovery led to one of the biggest debates in astronomy at the time.

- Some astronomers thought that nebulae were nearby clouds of dust, possibly areas of planetary formation.

- Other astronomers believed them to be distant galaxies so far away that they appeared fuzzy, this was known as the Island Universe Hypothesis.

The Curtis-Shapely debate

- The Curtis-Shapely debate was held in Washington in 1920 and revolved around two main astronomers.

- Heber Curtis
  - Curtis argued for the Island Universe Hypothesis.
  - He said that spiral nebulae were distant galaxies, just like our own, outside of our galaxy, which he thought was only 30 000 ly across.

- Harlow Shapely
  - Shapely thought our galaxy was much bigger – around 300 000 ly across, and formed the entire Universe.
  - He believed that spiral nebulae were nearby and just small clouds of gas.

- There was no conclusive evidence for either idea and it would be another five years until Edwin Hubble provided new evidence that showed that spiral nebulae were other galaxies that were much further away than astronomers previously thought possible.

- In the end neither Curtis nor Shapely was 100% correct.
  - Curtis was correct in that spiral nebulae were other galaxies but his estimate for the size of the Milky Way was far too small.
  - Shapley's estimate for the size of the Milky Way was closer but he was wrong about the nebulae.

> **Remember!**
> Not all nebulae are distant galaxies, some are clouds of dust, others clusters of stars, some are the remnants of an exploded star, and some very distant nebulae are clusters of galaxies.

Ideas about science

You should be able to:

- Describe how spiral nebulae were the main issues of the Curtis-Shapely debate, what each astronomer believed and what Hubble's evidence finally proved.

Improve your grade

The Curtis-Shapely debate

Foundation: Outline the key issue of the Curtis-Shapely debate and why at the time no agreement was reached.

AO1 [3 marks]

The expanding Universe

Edwin Hubble

- In 1925 Edwin Hubble finally showed that some nebulae were actually galaxies.

- He discovered a Cepheid variable star in the Andromeda nebula that had a period of 31 days, which meant that it should be very bright (see page 51).

- The star appeared to be very faint and when Hubble calculated how distant it was he found that it must be at least 1 million light years away, well outside of our galaxy; he concluded it must be in a separate galaxy.

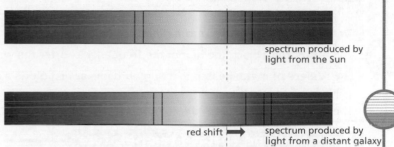

spectrum produced by light from the Sun

red shift ➡ spectrum produced by light from a distant galaxy

The dark lines in the spectrum from distant galaxies are shifted towards the red end of the spectrum. The bigger the shift, the faster the galaxies are moving apart.

- This also led to the realisation that the Universe was far bigger than anyone had imagined. We now know that the Universe is around 14 billion light years across and the Andromeda galaxy is actually one of the galaxies closest to us.

- When the spectrum of light from a galaxy is compared to the spectra produced by stars in our own galaxy it is found that the spectrum is shifted, almost always towards the red end of the spectrum.

- This **redshift** indicates that almost all galaxies are moving away from us.

- Hubble studied Cepheid variable stars in many distant galaxies and found that the further away a galaxy is the faster it is moving away from us.

- This discovery provided evidence for the Big Bang Theory, which

Hubble's original graph showing the link between recessional velocity (vertical axis) and distance (horizontal axis).

is believed by many scientists. It states that the entire Universe appeared at a single point about 14 thousand million years ago and has been expanding ever since

Hubble's Law

- Hubble's results form a straight line on a graph showing that recessional velocity is proportional to the distance from Earth. The constant of proportionality is called the **Hubble constant** such that:

  – Speed of recession (km/s) = Hubble constant ((km/s)/Mpc) × distance (Mpc)

  or

  – Speed of recession (km/s) = Hubble constant (s^{-1}) × distance (km)

- Values for the Hubble constant have been refined many times since the 1920s. Modern space telescopes have allowed study of Cepheid variable stars in distant galaxies, enabling us to calculate a better value for the Hubble constant. In March 2013 the Plank mission found the Hubble constant to be ~68 km/s per Mpc (mega parsec).

- The fact that almost every galaxy is moving away from us and the further away they are the faster they are moving away has led to the explanation that space itself is expanding.

- This expansion suggests that the Universe was much smaller in the past and so supports the Big Bang Theory.

> ### EXAM TIP
>
> You will need to be able to make calculations of the speed of recession using the above formula and for the higher paper you will need to rearrange the formula to find the Hubble constant or the distance. An easy way to do this is to use the triangle method.
>
> Speed of recession (Km/s)
>
> Hubble constant (Km/s Mpc) | Distance (Mpc)

Improve your grade

Hubble's Law

Foundation: Use Hubble's Law to calculate the speed of recession of a galaxy 15 Mpc away. Hubble's constant = 70 (km/s)/Mpc.

AO2 [2 marks]

The radiation from stars

Thermal radiation

- All objects emit a continuous range of electromagnetic radiation.
- The hotter the object, the more radiation it emits.
- The hotter the object, the higher the frequency (shorter the wavelength) of the emitted radiation.
- E.g. A piece of metal heated in a Bunsen flame at first glows a dull red and then gets brighter and whiter as it gets hotter.
- The same is true for stars – the colour and luminosity of a star depends on its surface temperature.

Radiation and surface temperature

- This graph shows how the luminosity and peak frequency of emitted radiation depends on a star's surface temperature. The curves show how much energy is emitted at each wavelength, with the total energy emitted being shown by the area under the curve.

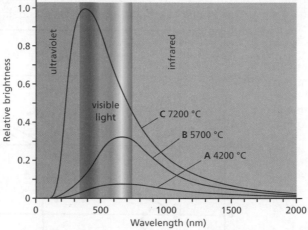

How the radiation emitted from a star depends on its surface temperture.

- Star **A** has a lower temperature and emits the least amount of energy, as shown by the area under the curve. It appears as an orange/red star, as shown by the highest point of the curve.
- Star **B** is a yellow star, like our Sun.
- Star **C** has the highest surface temperature, it is much more luminous, as shown by the area under the curve, and emits most of its radiation at the blue end of the spectrum so would appear blue/white. It also emits large amounts of ultraviolet radiation.

Energy transport in a star

- The centre of a star is where the density and temperature is highest; our Sun's core is around 16 million °C.
- This is where most of the nuclear fusion (see page 59) that powers the star takes place.
- These nuclear reactions release photons that travel outwards through the star, transferring energy as radiation through the **radiation zone**.
- Closer to the surface the star is cooler and energy is transferred by **convection** in the same way that convection works on Earth. Regions of stellar material deeper in the star are hotter so rise towards the surface where they transfer their energy, become cooler, and sink.
- Energy is radiated into space from the surface of the star (the **photosphere**).

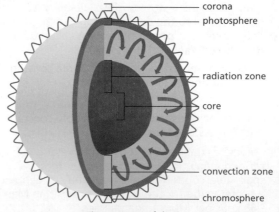

The structure of the Sun.

Improve your grade

Radiation and surface temperature

Foundation: the star Vega has a surface temperature of around 10 000 K while the star Procyon has a surface temperature of around 6000 K. Both stars have the same size. Compare Vega and Procyon in terms of brightness and colour.

AO2 [3 marks]

Analysing stellar spectra

Line spectra

- A rainbow is a continuous spectrum of colours covering all wavelengths of visible light.

- A **line spectrum** is where there are just a few coloured lines at specific wavelengths.

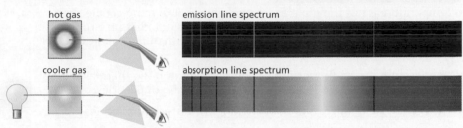

hot gas

cooler gas

emission line spectrum

absorption line spectrum

The black lines in an absorption spectrum exactly match the bright lines in an emission spectrum. These show the characteristic line spectra of hydrogen.

- Different elements produce different line spectra.

- Hot objects emit a spectrum of colours.

- If a gas is heated and the spectrum viewed through a prism it does not give a continuous spectrum but a line spectrum specific to the gas being heated. This is called its **emission spectrum**.

- If white light is shone through this gas some of the continuous spectrum is absorbed; this creates an **absorption spectrum** – the dark lines of the absorbed wavelengths exactly match the bright lines of the emission spectrum.

- Stars are very hot objects so emit a near continuous spectrum of radiation. However, if studied carefully it can be seen that there are hundreds of dark lines in their spectra.

- These dark lines are the result of gases in the cooler outer parts of the star absorbing certain wavelengths of light, with different elements absorbing different wavelengths and producing a characteristic spectrum.

The spectrum of the Sun – by careful analysis of the lines we can tell which elements are present.

- Studying these dark lines allows us to determine the chemical composition of the star.

- When a photon is absorbed by an atom the energy may be enough to remove an electron from the atom.

- This is called **ionisation**, with the atom that has lost an electron being called an ion.

- Different elements require different amounts of energy to ionise.

- The spectrum of the ion is different to that of the original atom.

- By looking at which ions are present we can determine how energetic the absorbed photons are and this gives an indication of the temperature.

Energy levels in atoms

- Atoms emit light when electrons lose energy.

- Line spectra indicate that electrons in an atom do not have a continuous range of energy otherwise the spectra would be continuous.

- Electrons in an atom exist in specific **energy levels**.

- A photon of light is emitted when an electron drops from one energy level to another. Different elements have their own set of allowed energy levels, so different elements have different emission spectra.

- The same idea explains absorption spectra; the dark lines occur where energy from the continuous spectrum has been absorbed and only light of the right energy to cause an electron to jump to a higher energy level can be absorbed.

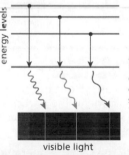

energy levels

visible light

As an electron jumps down between levels, it loses energy. This energy is emitted as a photon of light. Only certain energy levels are allowed, so only certain colours are emitted.

Higher

EXAM TIP

You will need to be able to identify the elements in a star based on information about its spectrum and line spectra of elements – simply look for which line spectra matches the dark lines in the star's spectrum. It may be that the line spectra of more than one element are needed to match all the dark lines in the star's spectrum.

Improve your grade

Comparing stars

Foundation: Stars much larger than the Sun are capable of producing heavier elements. How will their spectra be different to the Sun, which contains mainly hydrogen and helium?   *AO1, AO2* [3 marks]

Absolute zero

What is temperature?

- Temperature is not the same as thermal energy. A burning candle has a high temperature but not much energy and does not heat up a room. A radiator of hot water has a lower temperature than the candle but will warm a room because it has more energy.

- Temperature further can be thought of as how concentrated the thermal energy is and can be explained if we talk about the motion of particles.

- The higher the temperature of something the faster its molecules move.

- In a room the gas molecules are all moving around. As the temperature increases the faster the gas molecules move – this increases their kinetic energy.

- The temperature of a gas depends on the average kinetic energy of the molecules.

- If a gas is cooled the molecules lose kinetic energy and get slower. If the gas could be cooled low enough eventually the molecules would stop moving altogether; at this point the temperature could not get any lower.

- This coldest possible temperature is called **absolute zero**.

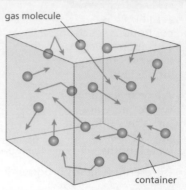
gas molecule

container

The molecules in a gas move in random directions at a range of speeds. The temperature depends on the average kinetic energy of the molecules.

Temperature and volume

- As the temperature of a gas increases the molecules move faster, causing the gas to expand and take up more space.

- If a volume of gas is trapped in a piston and the gas is heated, the molecules will move faster, hitting the piston with more energy and pushing the piston outwards, increasing the volume of the gas.

- If the gas is then cooled the molecules lose energy and the volume decreases again.

- If the gas continues to cool the volume will keep on reducing.

- Plotting a graph of temperature (°C) against volume produces a straight line graph which when extrapolated backwards reaches a point where the volume is zero at −273 °C.

- You can't have a negative volume so this is the coldest possible temperature (absolute zero).

- Starting the temperature scale at this point shows that volume is **directly proportional** to absolute temperature (measured in Kelvins (K)).

$V = 3.0 \text{ m}^3$

$V = 1.0 \text{ m}^3$

heat

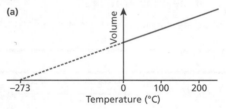

(a) Volume vs Temperature (°C): −273, 0, 100, 200

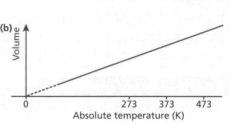

(b) Volume vs Absolute temperature (K): 0, 273, 373, 473

Temperature and pressure

- If a similar experiment was carried out but the volume was kept fixed, reducing temperature would mean the particles hit the wall of the container less and less frequently until eventually the molecules are not moving and the pressure falls to zero.

- This would occur at exactly the same temperature (absolute zero) and plotting a graph of temperature against pressure would be the same shape as the one above, indicating that pressure is directly proportional to temperature.

Remember!
When carrying out calculations involving gas volumes, temperatures and pressures you must work in Kelvins.

Remember!
For a fixed mass of gas:
Volume is proportional to temperature at constant pressure.
Pressure is proportional to temperature at constant volume.

Improve your grade

Collapsing cans

Foundation: A metal can is filled with hot air and then sealed. Explain why it begins to dent inwards after a few minutes.
AO1, AO2 [5 marks]

The gas laws

Compressing a gas

- Pressure in a gas is caused by gas molecules hitting the sides of the container.

- If the same number of molecules are put in a container of half the volume, they will hit the sides twice as often, doubling the pressure.

- So for a piston filled with a gas, increasing the volume reduces the pressure and reducing the volume increases the pressure.

- Changing the temperature or the amount of gas in the piston would affect this, so these are kept constant.

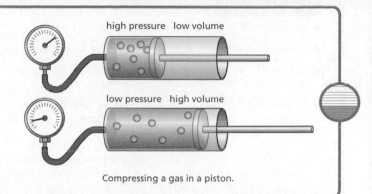

high pressure   low volume

low pressure   high volume

Compressing a gas in a piston.

Boyle's Law

- Robert Boyle was the first scientist to investigate this relationship and he formulated **Boyle's Law**.

- pressure × volume = constant

- This gas law basically says that as long as temperature and mass stay the same, if the pressure increases the volume must decrease and vice versa.

- The changes will always be consistent; if the pressure doubles, the volume halves, if the pressure is reduced by a factor of 10, the volume is increased by a factor of 10.

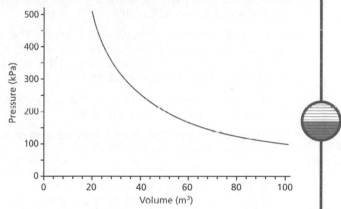

The graph shows how the pressure of a gas changes with its volume, if the temperature of the gas is kept constant and the mass of gas stays fixed.

Pressure, volume and temperature

- The state of a known mass of gas is described by three inter-related properties: pressure, volume and temperature.

- These quantities are linked by the three gas laws, which apply to a fixed mass of an 'ideal' gas.

- An ideal gas is one with well separated molecules that do not affect each other. A gas well above its boiling point and not under high pressure will behave as an ideal gas.

- The three gas laws are:

  – pressure × volume = constant

  – $\dfrac{\text{pressure}}{\text{temperature}}$ = constant

  – $\dfrac{\text{volume}}{\text{temperature}}$ = constant

- The laws can be combined such that $\dfrac{(\text{pressure} \times \text{volume})}{\text{temperature}}$ = constant.

- The first law (Boyle's Law) shows **inverse proportionality** – doubling one variable halves the other.

- The second two show **direct proportionality** – doubling one variable doubles the other.

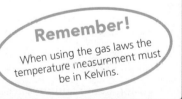

Remember!
When using the gas laws the temperature measurement must be in Kelvins.

Improve your grade

Scuba diving

Foundation: A cylinder of air is pressurised to 200 atmospheres at a temperature of 20°C. The cylinder is then left in the sun and heats up to 30°C. What will the new pressure be?   *AO1, AO2* [4 marks]

Star birth

Protostars

- Nebulae are immense clouds of molecular gas.

- They contain mainly hydrogen, helium and interstellar dust.

- Gravity is a force of attraction between all masses and this slowly pulls the gas and dust together.

- As the cloud collapses gravitational potential energy is transferred to kinetic energy, the particles move faster and the temperature increases.

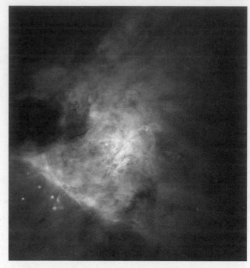

- Nebulae are not uniform, there are regions of higher density that are less stable and these clumps collapse into denser regions that will eventually become stars.

- These denser regions remain stable while the outward pressure is balanced by the inward gravitational attraction, but will sometimes collapse inwards once more.

- As they collapse inwards they become denser. The volume decreases, the gas particles move faster and so the temperature and pressure increases until the pressure and gravitational force are balanced again.

The *Orion* nebula is a massive cloud of gas with a large number of hot young stars.

- At this point the temperature has become so hot it glows red and is known as a protostar.

- Over the next few million years the protostar's gravity attracts more and more material to it.

- The more material it attracts the more massive it becomes and the stronger its gravitational attraction – this attracts even more material.

- Eventually the core becomes so hot and dense that nuclear fusion begins and a new star is born.

A computer simulation of a protostar.

Powering the stars

- For many years scientists debated about where the Sun got its energy from.

- It wasn't until the discovery of nuclear reactions in the early 20th century that scientists were able to provide a plausible explanation for where the energy came from and how the Sun could be so old (nearly 5 billion years) without running out of energy.

- Einstein's theory of relativity described how matter could be converted into energy and how the huge mass of the Sun is slowly being converted into energy.

- We now believe the Sun to be powered by **nuclear fusion**. Hydrogen nuclei are fused together to form helium and release huge amounts of energy in the process.

- Fusion in our Sun produces around 3×10^{26} Joules per second.

Improve your grade

Brown dwarfs

Foundation: Use your knowledge of physics to suggest why not all protostars become main sequence stars and smaller nebulae may produce brown dwarfs (failed stars), which are stars where hydrogen fusion has not begun.

AO2 [3 marks]

P7 Further physics – studying the Universe   Student book pages 170–171

Nuclear fusion

Fusion reactions

- The core of a star is where the temperature and density are highest; the core is many times denser than the densest material on Earth, but at over 15 million Kelvins the temperature is so high that it is still a gas.

- At these temperatures there are no atoms just a **plasma** of ions and electrons all whizzing about at incredibly high speeds.

- As most of a star is hydrogen, this means that most of the ions are protons and when these collide with each other at such high speeds they can fuse together. Whenever this occurs energy is released.

- The Sun is powered by the fusion of hydrogen into helium.

- Even in the Sun's core fusion reactions are so unlikely that an individual proton may hang around for a billion years before it fuses with another one.

- The high output of the Sun is due to its huge mass, not because fusion is easy.

- As helium has four nucleons and hydrogen only one, fusion into helium does not happen in one reaction but in a series of reactions called the **p-p cycle**.

- When writing nuclear reaction we do not worry about the outer electrons just the nucleus.

- We represent the nucleus in the following form $_Z^A X$ where A is the **mass number** (protons + neutrons), Z is the **atomic number** (the number of protons) and X is the chemical symbol.

- E.g. $_2^4 He$, a helium nuclei, has two protons and two neutrons so has a mass number of 4 and an atomic number of 2.

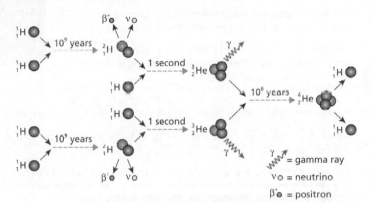

The p-p cycle is a chain of fusion reactions.

- Isotopes of an element have the same number of protons but different numbers of neutrons.

- The reaction of two protons fusing together is written as $_1^1 H + _1^1 H \rightarrow _1^2 H + _{+1}^0 e + _0^0 v$.
  - As we began with two positively charged hydrogen ions and are left with just one, a **positron** $_{+1}^0 e$ is produced in the reaction. Positrons also have the symbol β+ because they are a positively charged beta particle.
  - A **neutrino** $_0^0 v$ is also emitted. This has almost no mass and no charge.
  - Note that the mass and the atomic numbers on either side of the equation balance.

- The series of reactions that lead to a helium nucleus being formed involve a number of different **isotopes** of hydrogen and helium.

- In the diagram it can be seen that two hydrogen nuclei $_1^1 H$ fuse together to form **deuterium** $_1^2 H$ (an isotope of hydrogen).

- Deuterium then fuses with hydrogen to form helium-3 $_2^3 He$, which finally fuses with a second helium-3 nucleus to form helium−4 $_2^4 He$.

- In each of these reactions although the mass number is conserved mass is not, the total mass of the products is slightly less than the total mass of the reactants, the missing mass has been converted into energy.

Einstein's energy equation

- Einstein's energy equation $E = mc^2$ is used to calculate the amount of energy released when mass is converted into energy.
  - E = Energy, m = mass lost, c − speed of light in a vacuum

- In each fusion reaction a tiny amount of mass is converted into energy but there are around 10^{38} reactions per second in the Sun, amounting to the Sun losing around 3×10^9 kg per second.

- Applying Einstein's equation we find that fusion in the Sun produces $3 \times 10^9 \times (3 \times 10^8)^2 = 3 \times 10^{26}$ J/s.

Higher

Improve your grade

Fusion reactions

Foundation: Complete the fusion equation by inserting the missing numbers for the helium nucleus.

$_6^{12}C + _6^{12}C \rightarrow _8^{16}O + 2\,_{-}^{-}He + 2\,_0^1 n$

AO2 [2 marks]

The lives of stars

Main sequence stars

- Our Sun is a **main sequence** star.

- During the main sequence hydrogen is being fused into helium in the core and the outward pressure due to gas and radiation is balanced by the inward force of gravity.

- A star will remain in the main sequence fusing hydrogen into helium for most of its life, until it begins to run out of hydrogen.

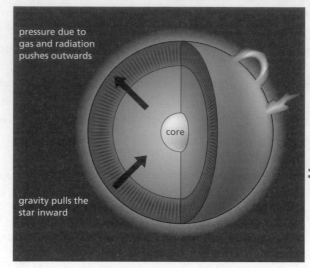

pressure due to gas and radiation pushes outwards

core

gravity pulls the star inward

A star in the main sequence is in equilibrium.

- Some stars are much larger than our Sun.

- In these stars core temperatures and pressures are so high that they can form heavier elements.

- Helium nuclei will fuse to form beryllium, carbon and oxygen and if the temperature is high enough this will continue up to iron.

- Fusing these lighter elements releases energy but to form elements heavier than iron requires energy to be provided.

- Once the core stops releasing energy the forces are no longer balanced and gravity will begin to crush the star.

- Heavier elements are only formed in the final stages of the life of the largest stars.

The Hertzsprung-Russell diagram

- Stars vary a great deal in size, luminosity and temperature.
- The **Hertzsprung-Russell diagram** shows the differences between stars.

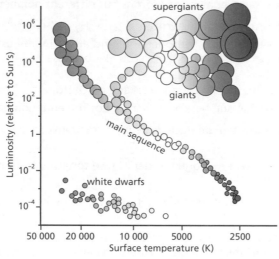

The Hertzsprung-Russell diagram.

- The luminosity (relative to the Sun) is plotted on the *y*-axis and the temperature is plotted on the *x*-axis.
- The stars on the Hertzsprung-Russell diagram fall into three main groups:
  - Main sequence stars, these form a diagonal line across the centre of the diagram. All of these stars are fusing hydrogen into helium but they vary greatly in temperature and luminosity.
  - **Giants** and **supergiants**, to the top right of the diagram. These are much larger than the Sun and are capable of fusing elements heavier than hydrogen. Some of these are coming to the end of their lives.
  - **White dwarfs**, to the bottom left. These are the small hot remnants of stars, nuclear fusion has ceased, they are gradually cooling and fading.

Improve your grade

Main sequence stars

Foundation: Use your understanding of physics to explain why a star becomes unstable and collapses towards the end of the main sequence. *AO2* [4 marks]

The death of a star

When the hydrogen runs out

- As a star runs out of hydrogen fusion is reduced, less energy is produced and the pressure drops.

- Gravitational attraction is now able to crush the star. At this stage the core is mainly helium surrounded by a shell of hydrogen.

- As the star collapses the density and temperature increase once more and fusion begins again. The energy released by this causes the star to expand enormously, its outer layer (the **photosphere**) cools as it expands and the star becomes a red giant or red supergiant.

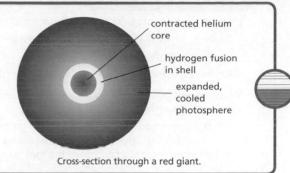

contracted helium core

hydrogen fusion in shell

expanded, cooled photosphere

Cross-section through a red giant.

The final fate of a star

- As the last of the hydrogen fuses more helium is produced. This gathers in the core, which becomes hotter and denser.

- When hot enough the helium begins to fuse into beryllium and then carbon.
  - $^{4}_{2}He + ^{4}_{2}He \longrightarrow ^{8}_{4}Be$
  - $^{4}_{2}He + ^{8}_{4}Be \longrightarrow ^{12}_{6}C$

- When low mass stars like the Sun become a red giant they don't have the mass to compress the core further at the end of helium fusion. At this stage fusion stops, the core collapses and the outer layers are ejected into space. The remnant is a slowly cooling white dwarf star surrounded by a **planetary nebula**. The white dwarf will eventually cool and fade to become a black dwarf.

- For stars with more than eight times the mass of the Sun things do not end with helium fusion – these super giants have enough mass that the gravitational attraction is able to fuse heavier elements. Carbon, nitrogen and oxygen and all elements up to and including iron are formed.

- When the core is mostly iron further fusion absorbs energy rather than releasing it, the star cannot generate any more energy so there is nothing opposing the gravitational attraction.

- At this point the star undergoes a catastrophic collapse called a **supernova**. The heaviest elements are formed during this collapse, which ends when the star explodes scattering material across space and leaving behind a small extremely dense core called a **neutron star**.

- With the largest stars the core is so dense that it collapses further to form a **black hole**.

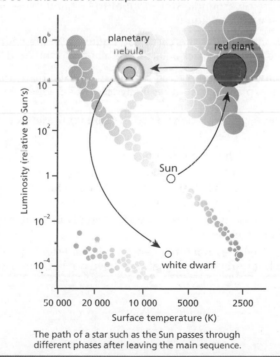

The path of a star such as the Sun passes through different phases after leaving the main sequence.

Improve your grade

Recycled stars

Foundation: The Sun is relatively small yet contains traces of amounts of iron. Use this information to make and justify a conclusion explaining where the material in the Sun came from. *AO2* [3 marks]

The possibility of extraterrestrial life

Extra-solar planets

- Finding planets around different stars is difficult.
  - They are too small to see at such a distance.
  - They don't give out their own light.
  - Light reflected from them is lost in the glare from the star.
- When looking for extra-solar planets astronomers study the star itself and look for signs of the affect an orbiting planet has on the star.
  - Wobble method – as a planet orbits a star the star is also attracted to the planet. Because the star is so much more massive than the planet the star does not move much but it wobbles enough that astronomers can detect tiny changes to its position or to the colour of its light caused by the Doppler effect.
  - Transit method – as a planet passes in front of a star it blocks a tiny amount of the light from the star so that the star's apparent brightness changes in a regular way as the planet orbits.

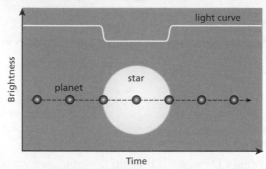

The transit method – light from the star dips in a regular way as the planet passes in front of it.

- Using these methods astronomers have found evidence of planets around hundreds of nearby stars.

The Goldilocks zone

- Most extra-solar planets discovered so far are Jupiter-sized planets orbiting close to the star as these have the biggest effect on the star. These planets are too close to the star to be habitable; they would be too hot and receive too much radiation.
- The Goldilocks zone is a region around a star thought to be habitable – neither too hot nor too cold and in a position where water would be liquid.
- Scientists know that very shortly after the Earth had liquid water life began, for this reason astronomers are particularly interested in finding rocky, Earth-like planets in the Goldilocks zone.
- As there are billions of stars in our galaxy and billions of galaxies many scientists believe that if only a small number of these have planets that life is likely to exist elsewhere in the Universe.

- Currently there is no generally accepted evidence that life (at present or in the past) exists anywhere other than on Earth.
- Scientists do have a few places they are looking:
  - Mars – there is evidence that Mars used to have liquid water so may once have had life. Some NASA scientists think some rocks they have studied show fossilised bacterial life but there are other possible explanations and this is not generally accepted.
  - Europa – one of Jupiter's moons has a frozen surface and liquid water oceans may exist below, there is the possibility of life but no evidence.
  - SETI – the Search for Extra-Terrestrial Intelligence project is listening for radio signals from space. Other than one unexplained but unrepeated signal they have not yet found anything.

Improve your grade

Life in our solar system

Foundation: Explain why scientists think that there may be life on Europa (one of Jupiter's moons).

AO2 [3 marks]

Observing the Universe

Space telescopes

- Space telescopes have a number of advantages and disadvantages when compared to ground-based telescopes.

- Advantages
  - They are not affected by absorption and refraction effects of the atmosphere.
  - They do not suffer from light pollution.
  - They are not affected by weather.
  - They can use parts of the electromagnetic spectrum that are absorbed by the atmosphere.

- Disadvantages
  - They are expensive to set up, maintain and repair.
  - Space programmes are very expensive so are not guaranteed to continue, e.g. the space shuttle has been retired so maintenance trips may now be more difficult and expensive to organise.

Ground-based telescopes

- The major optical and infrared astronomical observatories on Earth are mostly situated in Chile, Hawaii, Australia and the Canary Islands. The location of these telescopes is not an accident, the sites are chosen for a number of key reasons.
  - They are high above sea level so they are higher than most clouds and there is less atmosphere to affect the light before it reaches the telescope.
  - They have frequent cloudless nights.
  - The air is dry with low atmospheric pollution because water vapour and particles in the air would scatter the light, affecting the image.
  - They are a long way from built-up areas that cause light pollution.

Ground-based telescopes need to be at high altitude, where there are dry, cloudless nights, the air is free from particulate pollution and there is no light pollution.

- The remote location of these telescopes does present some difficulties.
  - Transport to the telescopes can take a long time and often new roads need to be built.
  - The need for new road systems and the distance from major built-up areas makes them expensive to build as all the materials need to be transported further.
  - Many astronomers are not able to make use of the telescopes as they are too far away.

Computer control

- Using a telescope is no longer about simply standing and looking through the eyepiece – much of modern astronomy is carried out by computer control, which has many important advantages.
  - Astronomers can log in to a telescope remotely so do not need to travel around the world to remote locations to use the best telescope, saving time and money.
  - The telescope can be programmed to continuously track objects or to scan and photograph specific regions of space in sequence.
  - The telescope can be positioned more precisely than by controlling it manually.
  - Images captured can be recorded and processed by the computer allowing analysis of wavelengths that are not visible to the human eye.

Improve your grade

Ground-based telescopes

Foundation: The Royal Observatory was built in 1676 on a hill in Greenwich Park, London, and played a major role in the history of astronomy. However it is no longer an active observatory. Suggest why the Royal Observatory was built in its original location and why it is no longer used.   *AO2* [5 marks]

International astronomy

Building a new telescope

- When building a new telescope there are a number of other factors that are considered in addition to how good the images will be. Some of these are:
  - Environmental and social impact near the telescope.
  - Working conditions for employees.
  - Cost.
- Astronomical telescopes are extremely expensive; the planned Extremely Large Telescope (ELT) will cost over €1 billion to build and about €50 million per year to run.
- By pooling the resources of several countries the cost of this telescope can be shared between the participating countries so that they all benefit from what will be the largest optical telescope ever built at a cost that they can afford.
- If each country built their own telescope no single telescope would be as large or as effective as the proposed ELT.
- Astronomers are also working together to plan and design the telescope, not only pooling financial resources but sharing expertise to achieve a better solution.

Working together

- Gamma ray bursts
  - Gamma ray bursts are extremely energetic bursts of gamma rays that come from space but no one is really sure what causes them.
  - They happen randomly and can last from a few seconds to several minutes, so observing them is difficult.
  - An international project is using satellites to detect the bursts. The Swift satellite is designed to detect the bursts then quickly send the coordinates to observatories across the world, which may be looking in the right direction to study them or the afterglow they leave behind.
- Supernova
  - The last supernova in our galaxy was over 400 years ago. Astronomers are waiting for the next one.
  - The first sign of a supernova will be a flood of neutrinos that will be detected at neutrino observatories, but they won't be able to pinpoint where the neutrinos came from.
  - They will send an alert to astronomers around the world who will then study the skies looking for the supernova. Once found they will share this information so that everyone can study this rare event.
- Online research
  - Telescopes and satellites now produce millions of digital images, far more than can be analysed by professional astronomers.
  - Computers can do some of the analysis but the human eye is better.
  - The Milky Way project is looking for bubbles of active star formation. By putting images from the Spitzer infrared space telescope online and by providing some simple software, researchers are getting the public to help them map our galaxy and learn more about the origin of stars.

Ideas about science

You should be able to:

- Describe two examples showing how international cooperation is essential for progress in astronomy.
- Explain non-astronomical factors that need to be considered when planning, building, operating and decommissioning an observatory.

Improve your grade

Building a new telescope

Foundation: Many new observatories are huge complexes, comprising many buildings and being operated by a large number of staff – some even require their own power station and increased road access. Evaluate possible advantages and disadvantages to explain if you think local people will be in favour of or opposed to a new observatory being built nearby.

AO1, AO2 [4 marks]

P7 Summary

The Sun, stars and the Moon appear to travel east-west across the sky. The Sun takes 24 hours, the stars slightly less and the Moon slightly more.

A sidereal day is the time it takes for the Earth to rotate 360 degrees. A sidereal day is 4 minutes less than a solar day due to the orbital movement of the Earth.

The positions of astronomical objects are described by angles of declination and angles of ascension.

A positive declination is above the equator, a negative declination below.

The angle of ascension tells you where to look east-west, usually called the angle of **right ascension**. It is measured in hours, minutes and seconds within 24 hours representing a full circle.

As the Moon orbits the Earth its appearance changes, this is due to how much of the lit side of the Moon we can see from Earth. During a full Moon we can see the entire lit side, during a new Moon we can only see the unlit face.

A lunar eclipse occurs when the Moon passes into the Earth's shadow and a solar eclipse occurs when the Moon's shadow falls on the Earth.

The planets appear to travel east-west across the sky but because they are also orbiting the Sun they sometimes appear to change speed and direction.

Naked eye astronomy

The speed of a wave is affected by the medium it is travelling through. When a wave travels from one medium to another its speed changes and this can cause it to change direction.

This direction change is called refraction and it occurs when a wave strikes the boundary at an angle other than 90 degrees.

The simple telescope uses two converging lenses, a low power objective lens to gather the light and a high power eyepiece lens to do the final magnification.

The objective lens should be as wide as possible to gather as much light as possible.

Magnification =
$$\frac{\text{focal length of objective (m)}}{\text{focal length of eyepiece (m)}}.$$

Diffraction occurs when a wave passes through a gap or around an obstacle.

The smaller the gap and the longer the wavelength, the more **effect** diffraction is.

Diffraction is worst when the wavelength is about the same size as the gap / aperture it is passing through.

In telescopes diffraction can reduce the resolution of the telescope. For this reason telescopes are made with as large an aperture as possible.

Light telescopes and images

A convex/converging lens uses refraction to form an image. Light entering near the edge of the lens is refracted the most and light entering near the middle the least.

Parallel rays of light entering the lens are brought to a focus at the focal point.

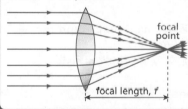

focal point

focal length, f

A reflecting telescope uses a large concave mirror instead of an objective lens.

A mirror can be made larger than a lens and so can see fainter objects.

A mirror does not suffer from chromatic aberration like a lens does.

When white light passes through a prism it can be split into the colours of the spectrum.

A diffraction grating can also split light into its spectrum.

A diffraction grating is either a series of thousands of very narrow gaps or thousands of lines drawn on a reflective surface. The lines / gaps are about the same width apart as the wavelength of light.

P7 Summary

Parallax is a method used to measure the distance to nearby stars. It is not possible to use parallax for more distant stars because the angles being measured are too small.

Distance (parsecs) =
$$\frac{1}{\text{parallax angle (seconds)}}$$

Luminosity of a star tells us the total amount of energy it gives off.

Luminosity depends on the size and temperature of the star.

A more distant star will appear less bright, the light it gives off spreads out as it travels through space.

Cepheid variable stars are stars that vary in brightness, their period tells you how rapidly they dim and brighten again.

Their luminosity depends on their period.

They are used to calculate the distance to distant galaxies.

It was the use of Cepheid variable stars to calculate distance that first proved that some nebulae were distant galaxies and not close-up dust clouds.

Mapping the Universe

The distance to other galaxies is measured in mega parsecs (Mpc).

Red-shift tells us that almost all galaxies are moving away from us.

Using Cepheid variable stars it was discovered that the faster the speed of recession, the further away the galaxy.

Speed of recession (km/s) = Hubble constant ((km/s)/Mpc) × distance (Mpc).

This suggests that the Universe is expanding, so used to be smaller. This supports the Big Bang Theory.

Absolute zero is the coldest possible temperature. At absolute zero particles stop moving completely so have no kinetic energy.

Absolute zero is −273 °C or 0 K.

When using the gas laws the temperature is measured in Kelvins.

If a fixed volume of gas is heated up the particles in it speed up, collide more often and the pressure increases.

Pressure/temperature = constant.

If gas at a fixed pressure is heated, it will expand.

Volume/temperature = constant.

If gas at a fixed temperature is compressed, the pressure will increase.

Pressure × volume = constant.

All hot objects emit a continuous spectrum of radiation.

The luminosity and peak frequency of the emitted radiation increases with temperature.

The movement of electrons between energy levels as they absorb photons of energy gives rise to line spectra.

Specific spectral lines in the spectrum of a star provide evidence of its chemical composition.

The Sun, the stars and their surroundings

A star is formed when clouds of dust and gas are pulled together by gravity until the temperature and pressure increase enough for nuclear fusion to begin.

A main sequence star fuses hydrogen into helium.

Towards the end of a star's life it will run out of hydrogen, it will start fusing helium and expand to become a red giant.

A red giant will fuse helium into heavier elements before running out of fuel and collapsing again to become a white dwarf.

The biggest stars will fuse heavier elements all the way up to iron before they explode as a supernova, creating the heaviest elements and then collapsing to become a neutron star or a black hole.

Nuclear fusion is when the nuclei of two atoms are joined together.

In a fusion reaction we represent the nucleus in the following form $^A_Z X$, where A is the mass number (protons + neutrons), Z is the atomic number (the number of protons) and X is the chemical symbol.

The reaction of two protons fusing together is written as $^1_1 H + ^1_1 H \longrightarrow ^2_1 H + ^0_{+1}e + ^0_0 v$. A positron $^0_{+1}e$ is produced in the reaction to ensure that charge is conserved.

Note that the mass and the atomic numbers on either side of the equation balance.

During fusion of light elements the total mass of the products is less than the total mass of reactant. The missing mass has been converted into energy according to the equation $E = mc^2$.

Astronomers have detected extra-solar planets by seeking the effect a planet has on its host star; an orbiting planet can make the host star wobble or dim slightly as it passes in front of it.

Scientists are looking for signs of life in places with liquid water and through radio signals but have not found any yet.

The astronomy community

Astronomers work together across the entire planet. This allows them to share resources and expertise as well as being able to share the workload of big projects.

Observatories are very expensive so countries share the cost of building them.

Ground-based telescopes need to be situated high up, in places with low atmospheric pollution, clear skies, dry atmosphere and away from cities to avoid light pollution.

Space telescopes have the advantages that they are not affected by the Earth's atmosphere and can use wavelengths that would be absorbed by the atmosphere, but they are very expensive to build, launch and maintain.

Page 4 The solar system

Foundation: Describe the motion of moons and planets in the solar system.   *AO1* [4 marks]

The planets go in circles around the Sun. Each planet is a ball of rock. There are moons around some of the planets. There are also smaller lumps called comets and asteroids. These also go around the Sun.

Answer grade: F. This answer is only worth 2 marks. It contains lots of information which was not asked for. For full marks you should say that moons orbit planets, and that the Sun is at the centre of the planet orbits.

Page 5 Fusion of elements in stars

Higher: Explain how most of the material in and around you was created by stars.   *AO1* [4 marks]

Stars give out energy because they can fuse hydrogen atoms into helium atoms. When they run out of hydrogen, the helium is fused to make heavier atoms such as carbon and oxygen. This needs a higher temperature and pressure in the star. In turn, these atoms then fuse to make even heavier atoms. This only happens in stars that are big enough for gravity to make the pressure and temperature high enough.

Answer grade: B. This answer is worth 3 marks. To gain full marks you would need to explain how the atoms get out of the star at the end of its lifetime – that it is the explosion of a star as a supernova that sends material into space where it can form a new solar system.

Many students assume that fusion can take place throughout a star, so all of the hydrogen has to be used up before helium fusion can start. In fact, fusion only takes place in the very middle. Even at the end of its lifetime, a star still contains a lot of hydrogen.

Page 6 Continental drift

Foundation: Explain why Wegener's theory of continental drift was not accepted when it was first published.   *AO1* [4 marks]

Wegener was not a geologist, so nobody paid attention to his ideas. He couldn't explain why the continents should move and their speed was too small to be measured with the instruments that they had then.

Answer grade: B. This answer sticks closely to the question and doesn't waste space by describing Wegener's theory. The three points made here are relevant and gain 1 mark each.

To push this answer to an A grade, you need to say a bit more about why geologists rejected Wegener's theory. You would need to explain that because geologists already had theories which explained a lot of Wegener's observations, they didn't see why they should accept his theory.

Page 7 Tectonic plates

Higher: Explain why there are volcanoes at plate boundaries.   *AO1* [4 marks]

Volcanoes allow rock to get out from the Earth's core onto the land. They can do this at plate boundaries because that is where the ground splits open to let the lava through.

Answer grade: F. This answer gains only 1 of the 4 marks. To get full marks, you need to discuss what happens when plates meet head-on and subduction occurs, as well as what happens where plates are moving apart. You also need to use the correct scientific terms, such as crust and mantle.

Page 8 Finding out about waves

Higher: Sound in steel has a speed of 2 km/s. What is the wavelength of a sound wave in steel which has a frequency of 80 000 Hz?   *AO2* [2 marks]

Speed = wavelength × frequency,

so $\text{wavelength} = \dfrac{\text{speed}}{\text{frequency}}$

$\text{Wavelength} = \dfrac{2}{80\,000} = 0.000025 \text{ m}$

Answer grade: B. The answer successfully rearranges the equation, but the student has forgotten to convert the speed into m/s before substituting it into the equation. For full marks, the answer should be 2000 / 80 000 = 0.025 m.

Many students calculate the wrong answer because they put the numbers straight into their calculator. If you write down the calculation first, then you would be able to do the sum again to check it.

Page 10 Ionisation

Higher: Bacteria are single-cell organisms that can pollute drinking water. Explain why exposing the water to ultraviolet light removes the bacteria, but exposing it to visible light does not affect the bacteria.

AO2 [4 marks]

The ultraviolet kills the bacteria because it is an ionising radiation. Light is not an ionising radiation, so does not kill the bacteria. Many bacteria absorb light to make their food, so light will probably make the pollution worse.

Answer grade: D/C. The first two sentences correctly explain why ultraviolet light removes bacteria from drinking water but visible light does not, and so gain 1 mark each. The last sentence gains no marks because it is not relevant to the question. To earn full marks, you would need to describe what happens to molecules when they are ionised (electrons are removed) and explain why only ultraviolet light can do this (its photons transfer enough energy).

Page 11 Microwaves

Higher: Explain the risks of cooking food with microwaves. *AO2* [3 marks]

Microwaves are strongly absorbed by water in our cells. This damages cells by ionising them, making them into cancer cells.

Answer grade: E. This is a well-ordered answer which states causes and effects in a logical order. However, there is an important factual error – microwaves are not an ionising radiation so are unlikely to lead to cancer. To get full marks you need to say that the microwaves transfer energy as heat in the water, and that too much heat will kill the cell.

Students often assume that all radiation damages people by giving them cancer. Microwaves and infrared don't have enough energy to ionise materials, so they transfer their energy as heat – and so can damage people by burning them.

Page 12 Global warming

Foundation: Explain how increasing carbon dioxide in the atmosphere results in global warming.

AO2 [4 marks]

The extra carbon dioxide stops infrared radiation from escaping into space. Global warming is going to melt all the ice and make the sea rise up and flood lots of land, making it difficult for us to grow all the food we need.

Answer grade: F. This answer gains 1 mark only – it has provided just one relevant scientific fact (sentence 1), but then wastes time describing some consequences of global warming. To get full marks you need to discuss the incoming energy from the Sun as light, as well as the outgoing radiation from the Earth as infrared.

Only answer the question you have been asked to! If you provide extra, unnecessary information, this will waste valuable time in an exam. This could cost you marks in later questions when time becomes short.

Page 13 Analogue and digital

Foundation: Describe the difference between analogue and digital signals used for radio broadcasts.

AO1 [4 marks]

Analogue sets the size of the wave, but digital turns it on and off. So an analogue signal looks like a wave which is getting gradually bigger and then smaller all the time. A digital wave is just there or not there.

Answer grade: D. This answer earns 3 marks without wasting any time in writing down unasked-for details (such as why the performance of digital is so much better than analogue). To earn full marks you would need to point out that digital performs better than analogue because it isn't affected so much by other signals such as noise and interference.

Page 15 Power

Foundation: A kettle comes with this warning:
This kettle must be used with a 230 V, 50 Hz supply. The current in the leads will be 8.7 A.

How much energy in kilowatt-hours is transferred from the supply when the kettle is used for 15 minutes?
AO2, AO3 [3 marks]

The power of the kettle = 230 V × 8.7 A = 2001 W

This is 2.001 kW.

The energy transferred = 2.001 × 15 = 30 kWh

Answer grade: E. The student has correctly calculated the power of the kettle, and earns 2 marks for this. They wisely selected the data required to do this, ignoring the 50 Hz.

To get full marks, you need to convert the 15 minutes into 0.25 hours for the second calculation. To help you avoid errors such as this, develop the habit of including the units as well as the numbers when you write down calculations.

Page 16 Efficiency

Foundation: Explain why governments have passed legislation which forces people to use energy-efficient lamps in their homes.
AO2 [4 marks]

Energy-efficient lamps transfer less energy wastefully to the environment than the old sort, for the same amount of energy transferred to light. This means that the nation needs to use less electricity.

Answer grade: D. The explanation of efficiency is excellent in this answer, and gains 3 of the 4 marks. To gain the final mark you would need to develop the argument to its logical conclusion, by explaining that using less electricity reduces our impact on the environment.

Page 17 How power stations work

Foundation: Describe how a power station transfers energy in coal to electricity.
AO1 [5 marks]

The coal is burnt to transfer heat energy to water, boiling it into high pressure steam. This is used to make a magnet rotate inside a wire, making electricity.

Answer grade: E. This answer earns 3 out of the 5 marks, one for each basic correct fact. However, one stage of the process (the turbine) is missing, and the description of the generator is incomplete (it should be a coil of wire). You would need to include both of these in your answer to gain full marks.

Page 18 Renewable energy sources

Foundation: The majority of the electricity in the UK is generated from fossil fuels. Explain the advantages and disadvantages of using wind and hydroelectric technology instead.
AO1 [4 marks]

Fossil fuels are non-renewable, so won't last forever. Wind and hydroelectric technologies are renewable energy sources, so can be used forever. Wind power has the problem that it only works while the wind is blowing and lots of people think that wind turbines are ugly. Hydroelectric power stations can only be built in mountains, a long way from where people want to use electricity. So perhaps we need both renewable and non-renewable sources for our electricity.

Answer grade: D. This answer correctly states the advantage of using wind and hydroelectric technologies to generate electricity. However, only the first disadvantage for wind technology earns a mark; the second disadvantage gains no marks because it is subjective and contains no science. Sentence 3 about hydroelectric power stations also earns no marks (distance is not a disadvantage because the National Grid transfers energy from faraway places to where it is needed). To gain full marks, you would need to state a disadvantage for the use of hydroelectric technology, such as its effect on the environment as it floods land or the expense of building it.

Page 19 Dealing with future energy demand

Foundation: It has been suggested that every person on Earth should only be allowed to produce a certain amount of carbon dioxide each year. Explain what impact this could have on your lifestyle. *AO2 [4 marks]*

I would have to use less energy from fossil fuels because these make smoke when they are burnt. I could use energy from renewable sources of energy such as solar and wind instead, but it might be difficult to make enough electricity this way.

Answer grade: D. This answer is quite good and gains 3 of the 4 marks. The candidate has stated three pieces of science, and has organised them to make a clear logical argument. Also, they have not wasted time talking about specific details, such as having to use a more efficient washing machine or only showering once a week, and have kept to general statements. To earn full marks you would need to mention carbon dioxide instead of smoke.

Page 21 Calculating speed

Higher: Henry is planning a train journey from Ipswich to Birmingham. The train journey is in two parts:

Departure time		Arrival time		Distance travelled (km)
12:00	Ipswich	13:00	Ely	80
13:15	Ely	15:45	Birmingham	120

a Which train is faster? Show your calculations.

b Taking into account the wait for the connection at Ely, what is the average speed of the journey from Ipswich to Birmingham?

a *The first train does 80 km in an hour. The second train does 120 km in an hour and a half, so they are both the same speed.*

b *The total time is 3.45 hours, so the average speed = 200 ÷ 3.45 = 58 km/h.*

Answer grade: D. The answer to part **a** is incorrect, as the students has said that the journey from Ely to Birmingham only takes 1½ hours not 2½ hours. In part **b** the student has said that 3 hours 45 minutes is 3.45 hours, not 3.75. However, the working is clear so some marks will be awarded.

Page 22 Speeding up

Higher: A cyclist rides around a velodrome track. When he sets off he takes 1 minute to reach a top speed of 14 m/s. Then he cycles 10 laps round the oval track at a steady speed. Calculate the cyclist's initial acceleration, and explain why he continues to accelerate after that.
AO1 [5 marks]

$$Acceleration = \frac{change\ in\ speed}{time} = \frac{14}{1} = 14\ m/s^2$$

He is still accelerating afterwards because his speed changes as he goes round the curves.

Answer grade: D/C. The student has used the correct equation to find acceleration, for which they gain 1 mark. However, they have forgotten to change 1 minute to 60 seconds. The correct calculation is 14 ÷ 60 = 0.23 m/s².

The explanation of why the cyclist continues to accelerate is accurate but incomplete. To gain more marks, you need to use the word velocity and explain that as the cyclist changes direction the direction changes, which means that the velocity changes.

Page 23 Forces between objects

Higher: James kicked his football towards the goal. There was a force on the ball when it was kicked. This force was part of an interaction pair.

Describe the partner force of the kicking force in the interaction pair.
AO1 [3 marks]

When James kicked the ball, he applied a force on the ball so it pushed back on his foot.

Answer grade: C/D. This answer is basically correct. When answering a question like this, make sure that it is clear that the force is *from* the football and on the foot. This answer is just about clear enough.

However, the student has forgotten to mention that the two forces in an interaction pair are equal and opposite. You would need to do this to get full marks.

Page 24 Terminal velocity

Higher: Explain how air bags reduce injury in a car crash. Include ideas about momentum in your answer.
AO1 [3 marks]

The air bag cushions the driver so he takes more time to stop. His momentum has reduced so the force is less.

Answer grade: C/D. This answer goes some way to explain how air bags reduce injury in a car crash, but is only partly correct. There is some confusion about change of momentum. The change of momentum does not depend on the time for the collision (only the mass and velocity change). To gain full marks, explain that the force is reduced because the change in momentum is slower.

Page 25 Energy transfers

Foundation: A spacecraft is returning to Earth. It has a gravitational potential energy of 8 MJ on re-entry.

a What is its maximum possible increase of kinetic energy as it falls?

b Explain why the actual increase of kinetic energy will be less than this value.
AO1 [3 marks]

a *8 MJ*

b *Because not all the potential energy becomes kinetic energy.*

Answer grade: F/E. The student has correctly calculated the answer to part **a**, and gains 1 mark for this. However, no marks are awarded for the answer to part **b**, because it does not explain *why*. For full marks, you need to say that the spacecraft slows down in the atmosphere due to air resistance.

P5 Improve your grade — Electric circuits

Page 27 Static electricity

Higher: Bella was rubbing a nylon comb with a duster. When she put the comb near his head, her hair moved towards the comb. Explain why this happened.

AO1 [5 marks]

When Bella rubbed the comb it got charged. When it went near her hair there was an electrostatic force of attraction so the hair was attracted towards the comb. The comb and the hair must have had opposite charge.

Answer grade: B/C. The first sentence does not explain how the comb becomes charged, so gains no marks. The second sentence makes good use of the key term *electrostatic force* and correctly describes the hair as being attracted towards the comb, gaining 2 marks. The student also correctly identifies the comb and hair has having opposite charge, and for this the answer gains an additional 1 mark.

For full marks, you would need to include a more detailed explanation about electrons being rubbed off the comb by friction, making the comb positively charged.

Page 28 Electrical resistance

Higher: Harry was recording values of current and voltage across a resistor so he could calculate the resistance.

a He recorded a current of 0.06 A when the voltage across the resistor was 4 V. Calculate the resistance.

b Harry then repeated the experiment with a voltage of 6 V. What current reading did he expect to get? Explain why he might not get this exact value.

AO1, AO2 [5 marks]

a *Resistance* $= \frac{4}{0.06} = 66.7$

b *Current* $= \frac{6}{66.7} = 0.09$ A

He might have connected it up wrong.

Answer grade: D/C. The student's answer to part **a** of this question gains 1 out of the 2 available marks for correctly calculating the resistance. To gain the second mark, you would need to include the unit for resistance (Ω).

In part **b**, the student has correctly calculated the current and has used the right unit, so gains 2 marks. However, the explanation about why the experiment might not give the exact value gains no marks. To gain full marks, you need to give reasons why there might be lower value for current, such as there might be a dirty connection or a broken lead.

Page 29 Series circuits and parallel circuits

Foundation: Susan and Mark were discussing adding resistors to a circuit. Susan said that if you added more resistors the total resistance would increase. Mark told her that sometimes adding more resistors would decrease the total resistance. Explain why both Susan and Mark are correct. *AO1* [5 marks]

When you add more resistors in series they are all in a line and you can add up all the resistors to give a bigger value for resistance, so Susan is right. Mark is talking about adding resistors in parallel – then the more resistors you add, the lower the total.

Answer grade: E/D. This answer is basically correct, giving a simple explanation of why both Susan and Mark are right, for which it gains 3 of the 5 marks available.

To gain full marks, the answer needs more detail. You would need to explain that adding resistors in series makes it harder for the charged particles to flow through the circuit. You also need to say that in parallel circuits the charged particles have a choice of pathway and it is easier to flow if there are more pathways.

Page 30 Generators

Foundation: Jenny uses a dynamo to power her bicycle lights.

a Explain why the lights are dim when she cycles slowly.

b Suggest one advantage and one disadvantage to using dynamo lights instead of battery powered lights. *AO1* [5 marks]

a *When you cycle slower the magnet doesn't spin as fast so there is a lower voltage.*

b *The advantage of using dynamo is that she doesn't need to buy batteries, and the disadvantage is that the light is dimmer.*

Answer grade: D. The student's answer to part **a** of this question is clear and accurate, gaining 2 of the 3 marks available. For full marks, you would need to give a more detailed explanation by saying that the lower induced voltage means lower current will flow through the bulb.

In part **b** of this question the student has correctly identified one advantage to using a dynamo, and gains 1 mark for this. However, a dynamo light is not necessarily dim. To gain full marks, you need to give a clear disadvantage, for instance, the light would go out when you had to stop to give way.

Page 31 Transformers

Higher: The mains supply at home is at 230 V a.c. A computer needs a supply at 23 V. Describe how the voltage of the mains is converted to the lower voltage. *AO1* [5 marks]

You need a step down transformer to reduce the voltage. The transformer has more coils on the primary coil and fewer on the secondary. The input voltage produces a magnetic field, which induces a lower voltage on the other coil.

Answer grade: C/B. This answer covers the basics and is accurate, so each sentence gains 1 mark.

For full marks, you would need to go into greater detail. First, you need to calculate the ratio of turns in the transformer (the primary coil needs 10 times as many turns as the secondary coil). Second, you need to say that the input creates a varying magnetic field to induce the output voltage.

Page 33 Radioactive elements

Foundation: Radioactive materials emit ionising radiation. Explain what is meant by 'ionising radiation'. *AO1* [3 marks]

Ionising radiation turns atoms into ions.

> **Answer grade: E/F.** While this answer is basically correct, it needs more detail. You need to say that when alpha or beta particles collide with an atom, some of the electrons are knocked off, leaving a positively charged ion.

Page 34 Hazards of ionising radiation

Higher: Explain why it is more dangerous to inhale (breathe in) an alpha emitter than a beta emitter. *AO1* [4 marks]

Alpha particles are more ionising, so they can cause a lot of cell damage, but they can't pass through skin.

> **Answer grade: C/B.** This answer is essentially correct, but if fails to explain why alpha radiation is more ionising than beta radiation. Also, while the last part of the answer is correct physics, it does not answer the question.
>
> To gain full marks, you need to say that an alpha emitter is more massive and has double the charge of a beta particle, so it is easier for it to ionise other atoms.

Page 35 Half-life

Higher: Radon-220 decays by alpha emission with a half-life of 52 seconds. The initial activity is 640 counts per second. How long will it take for the activity to become 80 counts per second? *AO1* [3 marks]

640 is 8 × 80 so it is 8 half-lives.

8 × 52 seconds = 416 seconds

> **Answer grade: C/D.** The first statement is correct (640 is 8 x 80), but the student has used the incorrect value to work out the time for the half-life. Half-life is the time it takes to halve the activity, so 640 – 320 – 160 – 80 is 3 half-lives. You would then multiply the half-life (52 seconds) by 3 to get the answer (156 s).

Page 36 Uses of ionising radiation

Higher: Radioactive tracers can be used in many different applications. One application is for research into plant nutrition.

Which of the following radioactive isotopes would be most suitable for studying plant nutrition? Explain your choice.

Isotope	Type of radiation	Half-life
Phosphorus-32	Beta decay	14 days
Nitrogen-16	Beta decay	7 seconds
Bismuth-210	Alpha decay	5 days

AO1 [3 marks]

The nitrogen, because it has a very short half-life so will not cause too much damage. It is also beta decay which is better than alpha, which would not be detected outside the plant.

> **Answer grade: B/C.** The arguments the student puts forward for beta decay are correct. However, the student has failed to think through the consequences of their decision – a half-life of 7 seconds is too short to study plant nutrition, as all the radioactivity will run out before the plant grows. For this reason, a beta emitter with a longer half-life (phosphorus-32) would be more suitable.

Page 37 Energy from the nucleus

Foundation: Explain the difference between nuclear fission and nuclear fusion. *AO1* [4 marks]

Nuclear fission is when an atom splits up and fusion is when two atoms combine to form a new atom.

> **Answer grade: E/F.** This answer is basic and accurate, so gains 2 marks. Note that there are 4 marks available for this question, so to gain full marks you need to give a more detailed response. You would need to explain that fission occurs with large unstable nuclei, such as uranium, and that fusion occurs with small nuclei such as hydrogen.

Page 39 Sidereal days

Higher: Explain the difference between the length of a solar day and the length of a sidereal day.

AO1 [4 marks]

A sidereal day is the time it takes for the Earth to rotate 360 degrees and is slightly shorter than a solar day, which is the time it takes for the Sun to appear in the same position from one day to the next.

Answer grade: C. The answer describes the difference but does not explain it and would receive 2 marks. For full marks you need to add that a solar day is longer because the Earth is also orbiting the Sun so the Earth needs to rotate just over 360 degrees for the sun to appear in the same position in the sky.

Page 40 The phases of the Moon

Foundation: During a new Moon the Moon is not visible in the night sky. What causes this?   *AO1 [2 marks]*

This is because the Earth is between the Moon and the Sun so the Moon is in the Earth's shadow.

Answer grade: U. This answer is incorrect; it is a common mistake and actually describes a lunar eclipse. For full marks you would need to say that a new Moon occurs when the Moon is on the same side of the Earth as the Sun and we can only see the unlit (dark side) of the Moon.

Page 41 Planets

Foundation: Although smaller than Jupiter, the planet Venus appears brighter. Suggest why.   *AO2, [2 marks]*

It is closer to the Earth and to the Sun than Jupiter.

Answer grade: D. The answer is correct but only scores 1 mark. For full marks the answer should also mention that it is more reflective or expand on the fact that it is closer, meaning that light reflected from it does not spread out as far by the time it reaches us.

Page 42 Using angles of ascension and declination

Foundation: Compare the positions of the two stars described in the table below.   *AO2, AO3 [3 marks]*

	Angle of right ascension	Angle of declination
Star A	30 degrees	−10 degrees
Star B	30 degrees	+75 degrees

Both stars are in the same position east-west. Star A is south of the equator and star B is north of the equator.

Answer grade: B. The answer scores 2 marks because it has covered ascension and declination correctly and refers to both the things being compared. For full marks you need to add that star A is **just south** of the equator while star B is a **long way north** of the equator.

Page 43 Refraction

Foundation: Light travels slower in water than in air; use this information to complete the diagram by adding ray lines to show how the observer is able to see the bottom of the cup.

AO2 [3 marks]

Example answer:

Answer grade: D. Answer scores 1 mark for using straight lines. A common mistake has been made with the direction change. Light entering or leaving a surface at right angles (along the normal) is not refracted – the ray needs to be drawn coming out at an angle and for full marks one arrow needs to be added to each ray showing the light travelling from the coin to the eye.

Page 44 The power of a lens

Foundation: A magnifying glass has a focal length of 10 cm. Calculate the power of the lens.   *AO2 [3 marks]*

$$Power = \frac{1}{focal\ length} = \frac{1}{10} = 0.1D$$

Answer grade: C. Answer scores 2 marks. The candidate selected the correct formula and remembered to add the units for power but needed to convert the focal length to metres before doing the calculation. Should have been $\frac{1}{0.1} = 10D$.

Page 45 Magnification

Higher: A refracting telescope has an eyepiece lens with a focal length of 5 cm and an objective lens with a magnification of 1 m. Find its magnification and explain what this means.   *AO2 [4 marks]*

$$Magnification = \frac{focal\ length\ of\ objective}{focal\ length\ of\ eyepiece} = \frac{1}{0.05} = 20X$$

The image will look 20 times bigger.

Answer grade: B. Answer scores 3 marks. For full marks the answer needs to be more specific. The image will be 20 times bigger *than if viewed with the naked eye.*

Page 46 Why use mirrors?

Foundation: Reflecting telescopes are not affected by chromatic aberration. Explain another reason why most astronomical telescopes are reflecting telescopes instead of refracting telescopes.

AO1, AO2 [3 marks]

Reflecting telescopes give better images because they can be made bigger.

Answer grade: E. Answer lacks detail and would only score 1 mark. Questions need to not only describe the advantages but explain them. For full marks the answer needs to include that having a bigger objective lens means that reflecting telescopes can collect more light, which allows fainter objects to be seen.

Page 47 Diffraction and telescopes

Higher: The table below gives some information about telescopes. Use this information to put the telescopes in order of how much they are affected by diffraction, from the most affected to the least affected.

AO2, AO3 [3 marks]

Telescope	Radiation detected	Wavelength detected	Aperture size
James Clerk Maxwell	Microwave	1 m	15 m
Proposed ELT	Visible light	5×10^{-7} m	42 m
Hopkins	Ultraviolet	1×10^{-7} m	1 m
Arecibo	Radio waves	100 m	300 m

Hopkins, James Clerk Maxwell, Proposed ELT, Arecibo.

Answer grade: D. Only one telescope is in the correct place, the answer has just been based on aperture size – this is correct if all were looking at the same wavelength but they are not, the amount of diffraction depends on aperture and wavelength. You need to look at how many times bigger than the wavelength the aperture is. The greater this number, the less diffraction takes place. So the answer should be: *Arecibo, James Clerk Maxwell, Hopkins, Proposed ELT.*

Page 48 Looking at evidence

Foundation: Just after Isaac Newton earned his maths degree at university he was playing with prisms at home and saw that a different coloured light came out to that which went in. He proposed that white light was a mixture of different colours. Suggest why Newton's ideas were not generally accepted straight away. *AO2 [3 marks]*

Newton's ideas were not accepted because he was a mathematician and not a recognised scientist at the time.

Answer grade: C. The answer only makes one point, for 3 marks the answer should suggest three points. These could include any of the following:

He did not have much evidence to support his ideas.

Other scientists had not tested and verified his results.

His results could be explained in other ways, e.g. the prism changed the colour of the light.

He had not made and tested predictions based on his ideas.

Page 49 Parallax

Foundation: Explain why parallax can only be used to measure the distance to nearby stars. *AO2 [2 marks]*

The further away the star, the smaller the parallax angle.

Answer grade: C. The answer is true but incomplete. For full marks you should add that when too far away it is impossible to measure the parallax angle accurately because the angle is too small.

Page 50 Luminosity and apparent brightness

Foundation: The table shows three well-known stars. Use the data in the table to order the stars according to which will appear brightest from the Earth, starting with the brightest. *AO2, AO3 [2 marks]*

Star	Luminosity	Distance from Earth
Alpha Centauri	$1.5\ L_o$	4.3 ly
Sirius	$25.4\ L_o$	8.6 ly
Rigel	$126{,}000\ L_o$	860 ly

Rigel, Sirius, Alpha Centauri.

Answer grade: D. Answer scores 1 mark. Correct answer is: Sirius, Rigel, Alpha Centauri. Rigel is 5000 times more luminous than Sirius but it is 100 times further away so its light is reduced by 100×100 or 10,000 times, meaning it appears half as dim. Alpha Centauri is closer than Sirius but over 10 times less luminous.

Page 51 Variable stars

Higher: An astronomer observes two Cepheid variable stars, one with a period of 12 days and one with a period of 20 days.

a Which star is most luminous?

b Explain how this can help the astronomer find the distance to the stars and what other information is needed. *AO2, AO3 [4 marks]*

a *The star with a period of 20 days is most luminous.*

b *To find the distance the astronomer needs to look at how bright the stars appear from Earth – the brighter it is the closer it is.*

Answer grade: C. Part a is correct, part b is half right. The astronomer needs to look at how bright the stars appear but then needs to compare this to their luminosity to work out the distance – it is not always the case that the brighter the star the closer it is.

Page 52 The Curtis-Shapely debate

Foundation: Outline the key issue of the Curtis-Shapely debate and why at the time no agreement was reached. *AO1 [3 marks]*

The key issue was whether or not spiral nebulae were distant galaxies or nearby dust clouds, no agreement was reached because there was not enough evidence either way.

Answer grade: C. Answer scores 2 out of 3. For full marks need to add *and the evidence there was could be interpreted in different ways.*

Page 53 Hubble's Law

Foundation: Use Hubble's Law to calculate the speed of recession of a galaxy 15 Mpc away.

Hubble constant = $\dfrac{70 \ (km/s)}{Mpc}$.   *AO2 [2 marks]*

Speed of recession = 70 × 15 = 1050.

Answer grade: D. Answer is correct but unit is missing, for full marks should be: 1050 km/s.

Page 54 Radiation and surface temperature

Foundation: The star Vega has a surface temperature of around 10 000K while the star Procyon has a surface temperature of around 6000 K. Both stars have the same size. Compare Vega and Procyon in terms of brightness and colour.   *AO2 [3 marks]*

It is more luminous and will give out light of a higher frequency. It will therefore be brighter and bluer. How bright it appears from Earth will also depend on how far away it is.

Answer grade: U. Answer is 100% correct, however the answer just says 'it' so could also be 100% incorrect and scores nothing – because the examiner does not know which star is being talked about. The answer should have been: *Vega is more luminous and will give out light of a higher frequency than Procyon. Vega will therefore be brighter and bluer than Procyon but which star appears brightest from Earth will also depend on how far away they are.*

Page 55 Comparing stars

Foundation: Stars much larger than the Sun are capable of producing heavier elements. How will their spectra be different to the Sun, which contains mainly hydrogen and helium?   *AO1, AO2 [3 marks]*

The spectra of the Sun will have dark lines showing hydrogen and helium whereas a larger star will have dark lines showing the heavier elements.

Answer grade: C. Answer scores 2 marks. For full marks the answer should have said that the larger star will have dark lines showing the heavy elements *as well as dark lines showing hydrogen and helium.*

Page 56 Collapsing cans

Foundation: A metal can is filled with hot air and then sealed. Explain why it begins to dent inwards after a few minutes.   *AO1, AO2 [5 marks]*

The can is a fixed volume and contains a fixed mass of gas, as it cools down the pressure drops until it is so low that the air pressure outside crushes the can.

Answer grade: C. The answer scores 3 marks. For 5 marks a detailed answer is required. For full marks the answer should have included the kinetic model to explain why the pressure drops. *As the gas cools the particles move slower, lose energy, hit the walls of the can less often and therefore the pressure drops...*

Page 57 Scuba diving

Foundation: A cylinder of air is pressurised to 200 atmospheres at a temperature of 20 °C. The cylinder is then left in the sun and heats up to 30 °C. What will the new pressure be?   *AO1, AO2 [4 marks]*

$\dfrac{Pressure}{temperature} = constant$

$\dfrac{200}{20} = \dfrac{new \ pressure}{30}$

$New \ pressure = \left(\dfrac{200}{20}\right) \times 30 = 300 \ atmospheres$

Answer grade: D. Answer scores 2 out of 4. The student forgot to convert °C into Kelvins first so the answer is incorrect, however showing the working out meant that method marks were awarded – if working had not been shown and just the final answer of 300 atm had been given, the student would have scored zero. For full marks the answer should have been:

20 °C – 293 K, 30 °C – 303 K.

$New \ pressure = \left(\dfrac{200}{293}\right) \times 303 = 206.8 \ atm.$

Page 58 Brown dwarfs

Foundation: Use your knowledge of physics to suggest why not all protostars become main sequence stars and smaller nebulae may produce brown dwarfs (failed stars), which are stars where hydrogen fusion has not begun.   *AO2 [3 marks]*

A brown dwarf could be where the dust cloud and protostar is not big enough for fusion to start.

Answer grade: U. The answer makes the exam technique error of simply restating and rewording the information given in the question so scores zero marks. For full marks the candidate should have applied understanding and said, *when there is not enough material the protostar's mass and gravity; is too low and therefore temperature; and pressure; are too low for fusion to begin.*

Page 59 Fusion reactions

Foundation: Complete the fusion equation by inserting the missing numbers for the helium nucleus.   *AO2 [2 marks]*

$^{12}_{6}C + ^{12}_{6}C \longrightarrow ^{16}_{8}O + 2^{-}He + 2^{1}_{0}n$

$^{12}_{6}C + ^{12}_{6}C \longrightarrow ^{16}_{8}O + 2^{4}_{2}He + 2^{1}_{0}n$

Answer grade: D. The candidate has remembered and inserted the most common isotope, helium-4. This balances the atomic numbers correctly but does not balance mass numbers in the equation so would score 1 mark. The candidate should have made sure the mass and atomic numbers on either side balance, doing so would have given the following:

$^{12}_{6}C + ^{12}_{6}C \longrightarrow ^{16}_{8}O + 2^{3}_{2}He + 2^{1}_{0}n$

Page 60 Main sequence stars

Foundation: Use your understanding of physics to explain why a star becomes unstable and collapses towards the end of the main sequence. *AO2* [4 marks]

As the star runs out of fuel it produces less energy so the force of gravity acting inwards is greater than outward forces and the star collapses.

Answer grade: D. The answer scores 2 marks. For full marks the answer needs to explain that as the star produces less energy its temperature falls and so the pressure falls, reducing the outward force.

Page 61 Recycled stars

Foundation: The Sun is relatively small yet contains traces of amounts of iron. Use this information to make and justify a conclusion explaining where the material in the Sun came from. *AO2* [3 marks]

The material in the Sun must have come from a large star that produced the iron and then exploded in a supernova.

Answer grade: C. The conclusion is correct and would score 2 marks but it is not fully justified since it does not explain why the Sun could not have made its own iron. For full marks the candidate needs to say that the Sun is too small to produce iron so the iron must have been produced in a large star, which then exploded in a supernova.

Page 62 Life in our solar system

Foundation: Explain why scientists think that there may be life on Europa (one of Jupiter's moons). *AO2* [3 marks]

Because there might be liquid water there.

Answer grade: E. The answer does not explain why liquid water is important and only scores 1 mark. For a 3-mark question try and give three points. For full marks the answer needs to add that all life on Earth requires liquid water and that life appeared on Earth very soon after we had liquid water.

Page 63 Ground-based telescopes

Foundation: The Royal Observatory was built in 1676 on a hill in Greenwich Park, London, and played a major role in the history of astronomy. However it is no longer an active observatory. Suggest why the Royal Observatory was built in its original location and why it is no longer used. *AO2* [5 marks]

It was built in London because that's where lots of people lived and travel was not as easy in 1676. It is not used any more because all the lights of London cause light pollution and England has lots of clouds.

Answer grade: D. This is a 5-mark question so the candidate should choose five points related to the siting of ground-based telescopes – the answer has only three points: transport, light pollution, and cloud cover. The answer should include two further points from: *atmospheric pollution, altitude, cost, and humidity.*

Page 64 Building a new telescope

Foundation: Many new observatories are huge complexes, comprising many buildings and being operated by a large number of staff – some even require their own power station and increased road access. Evaluate possible advantages and disadvantages to explain if you think local people will be in favour of or opposed to a new observatory being built nearby. *AO1, AO2* [4 marks]

I think people will be in favour because it will bring jobs to the area, the observatory staff will also spend money locally and the roads will be upgraded, making travel easier.

Answer grade: B. No mark is given for in favour or against. Marks are for a reasoned argument. The answer made three points and scored 3 marks. For full marks the candidate could have added: even though there will be more traffic the roads will be upgraded; the observatory could bring additional support industries. An argument against could cover: more traffic so accompanying noise and pollution; environmental impact (traffic noise, pollution, visual, habitat damage); cost if taxpayer funded; increased drain on local resources, e.g. education, healthcare.

Ideas About Science

Understanding the scientific process

As part of your Science assessment, you will need to show that you have an understanding of the scientific process – Ideas about Science.

Science aims to develop explanations for what we observe in the world around us. These explanations must be based on scientific evidence, rather than just opinion. Scientists therefore carry out experiments to test their ideas and to develop theories. The way in which scientific data is collected and analysed is crucial to the scientific process. Scientists are sceptical about claims that cannot be reproduced by others.

You should be aware that there are some questions that science cannot currently answer and some that science cannot address.

Collecting and evaluating data

You should be able to devise a plan that will answer a scientific question or solve a scientific problem. In doing so, you will need to collect and use data from both primary and secondary sources. Primary data is data you collect from your experiments and surveys, or by interviewing people.

While collecting primary data, you will need to show that you can identify risks and work safely. It is important that you work accurately and that when you repeat an experiment, you get similar results.

Secondary data is found by research, often using ICT (the Internet and computer simulations), but do not forget that books, journals, magazines and newspapers can also be excellent sources. You will need to judge the reliability of the source of information and also the quality of any data that may be presented.

Presenting and processing information

You should be able to present your information in an appropriate, scientific manner, using clear English and the correct scientific terminology and conventions. You will often process data by carrying out calculations, drawing a graph or using statistics. This will help to show relationships in the data you have collected.

You should be able to develop an argument and come to a conclusion based on analysis of the data you collect, along with your scientific knowledge and understanding. Bear in mind that it may be important to use both quantitative and qualitative arguments.

You must also evaluate the data you collect and how its quality may limit the conclusions you can draw. Remember that a correlation between a factor that's tested or investigated and an outcome does not necessarily mean that the factor caused the outcome.

Changing ideas and explanations

Many of today's scientific and technological developments have benefits, risks and unintended consequences.

The decisions that scientists make will often raise a combination of ethical, environmental, social and economic questions. Scientific ideas and explanations may change as time passes, and the standards and values of society may also change. It is the job of scientists to discuss and evaluate these changing ideas, and to make or suggest changes that benefit people.

Glossary

A

absolute zero the temperature at which all molecular movement stops; equivalent to –273 °C 56

absorb to take in energy from electromagnetic radiation; this is transferred to the particles of the material 10, 11, 12, 14

absorption spectrum A series of dark lines on a continuous spectrum, caused by the absorption of photons of specific wavelengths; can be caused by the cooler gases surrounding a star 55

acceleration the rate at which the velocity of an object changes 22, 24, 26

activity (radioactivity) the amount of radiation emitted from a material 35, 38

air resistance the upwards force exerted by air molecules on an object 24–25

alpha particles (α) radioactive particles which are helium nuclei – helium atoms without the electrons (they have a positive charge) 33–34

alternating current (a.c.) an electrical current in which the direction of the current changes at regular intervals 30–32

ammeter a device that measures the amount of current running through a circuit in Ampères 28

Ampères (amps) the unit of measurement used for the flow of electrical current or charge 15, 27

amplitude the maximum disturbance of a wave motion from its undisturbed position 8–9, 13

analogue equipment that can display data with continuous values 13–14

analogue signal transmitted data that can have any value 13–14

aperture opening at the front of a telescope (or the eye) through which light can enter 47

asteroid small object in orbit in the solar system 4, 9

atmosphere thin layer of gas surrounding a planet 11–14

atom the basic 'building block' of an element which cannot be chemically broken down 5, 10, 33–37

atomic number the total number of protons in a nucleus 59

attractive a force that pulls two objects together 23

average speed distance travelled divided by the time taken 21, 26

B

background radiation low-level radiation that is found all around us 23, 33, 38

bacteria single-celled microorganisms, some of which may invade the body and cause disease 10, 36

beta particles (β) particles given off by some radioactive materials (they have a negative charge) 34–35, 38

Big Bang the theoretical beginning of the Universe, when energy and matter expanded outwards from a point 5

binary digit a number that can only take the values 0 or 1 13

binding energy the energy that holds particles together in a nucleus 36, 38

black hole a region of enormous gravitational force, sometimes due to the collapse of a supergiant star 61

Boyle's law the pressure of a fixed mass of an ideal gas is inversely proportional to its volume, at a constant temperature 57

byte a measure of digital data consisting of 8 binary digits 13–14

C

carbon an element that combines with others, such as hydrogen and oxygen, to form many compounds in living organisms 12, 14–15, 19–20, 34–35, 37

carbon cycle the way in which carbon atoms pass between living organisms and their environment 12

carbon dioxide gas whose molecules consist of one carbon and two oxygen atoms, CO_2; product of respiration and combustion; used in photosynthesis; a greenhouse gas 12, 14–15, 19–20, 37

carrier wave electromagnetic wave on which a signal is superimposed for transmission 12

celestial sphere an imaginary sphere of large radius centred on the Earth; all the objects in the sky are thought of as being on the sphere. The celestial sphere rotates with the Earth 42

Cepheid variable stars a type of star with luminosity that varies in a regular way; the period of the variation depends on the size of the star 51

chain reaction a fission reaction that is maintained because the neutrons produced in the fission of one nucleus are available to initiate fission in other nuclei causing a rapid production of energy 37–38

chromatic aberration coloured fringes seen on an image due to different refraction of different wavelengths of light 46

comet lump of rock and ice in a highly elongated orbit around the Sun 4, 9

components devices such as lamps and motors on an electrical circuit to which energy is transferred 16, 27–29, 32

conductor a substance in which electric current can flow freely 27–29, 32

conservation of energy when energy cannot be created or destroyed 25

contaminated having mixed with something harmful such as a pollutant or radioactive substance 18, 37

contamination (radioactivity) something that comes into contact with radioactive material 18–19, 20, 34, 38

continental drift slow movement of continents (land masses) relative to each other 6, 9

control rods absorb excess neutrons in order to control a chain reaction 37–38

convection heat transfer in a liquid or gas, when particles in a warmer region gain energy and move into cooler regions, carrying this energy with them 9, 12, 17, 54

converge bring together; a converging lens refracts light rays so as to bring them to a focus 43

convex lens a lens that is thicker in the middle than at the edges 43, 44

coolant gas or liquid that circulates around a reactor to keep it cool 37–38

correlation a link between two factors that shows they are related, but one does not necessarily cause the other; a positive correlation shows that as one variable increases, the other also increases; a negative correlation shows that as one variable increases, the other decreases 11–12, 14

crust surface layer of Earth, made up of tectonic plates 6–7, 9

current flow of electrons in an electric circuit 9, 15–18, 20, 26–32

D

data information, often in the form of numbers obtained from surveys or experiments 4, 8–9, 12

daughter product the name given to the radioactive element formed from the decayed initial radioactive element 35

decay chain a series of radioactive decays of an unstable nucleus to the nucleus of a different element, until a stable nucleus is formed 35

declination one of two angles that describe a star's location on the celestial sphere. Declination is similar to latitude and gives the star's position in degrees north or south of the celestial equator 42

decode to extract information from a code 13

deuterium a heavier isotope of hydrogen with one neutron in the nucleus 59

diffraction the spreading out of waves round an obstacle or through a gap 47

Glossary

diffraction gratings set of ruled lines through which light is transmitted, or reflected; used for creating spectra 48

digital signal transmitted information that can take only a small number of discrete values, usually just 0 and 1 13, 14

dioptres unit of refractive power, abbreviated to D; a lens of power 1 D has a focal length of 1 m 44

direct current (d.c.) an electric current that flows in one direction only 30, 32

directly proportional two variables are directly proportional if their ratio is constant, for example, if one quantity doubles, the other one also doubles. A graph of the two variables would be a straight line through the origin 56, 57

displacement the distance moved in a specific direction 21, 26

displacement–time graph a visual way of showing the displacement (distance and direction from a starting point) of an object against time 21

distance–time graph a visual way of showing the time taken for a journey and the distance travelled 21, 26

DNA large (polymer) molecule found in the nucleus of all body cells – its sequence determines genetic characteristics, such as eye colour, and gives each one of us a unique genetic code 34

drag see air resistance 24, 26

dwarf planet spherical object orbiting the Sun, smaller than a planet and larger than an asteroid 4, 9

E

ecliptic a line on the celestial sphere that shows the apparent path of the Sun relative to the stars 42

efficiency a measure of how effectively an appliance transfers the input energy into useful energy 16, 18–20

electric current a negative flow of electrical charge through a medium, carried by electrons in a conductor 15, 17, 20, 27, 30–32

electromagnetic induction a term used by Faraday to explain induced voltage 30, 32

electromagnetic radiation energy transferred as electromagnetic waves 10, 12, 14

electromagnetic spectrum electromagnetic waves ordered according to wavelength and frequency – ranging from radio waves to gamma rays 10

electron tiny negatively charged particle within an atom that orbits the nucleus – responsible for current in electrical circuits 10, 27–29, 32–34, 37–38

electrostatic force a force caused by positive and negative charges 27

element substance made out of only one type of atom 5, 33, 35, 38

emission spectrum a line spectrum from a hot vapour 55

energy input the energy transferred into a device or appliance from elsewhere 16, 20

energy levels The allowed orbits for an electron in an atom 55

energy output the energy transferred away from a device or appliance, which may be either useful or wasted 16

environment an organism's surroundings 16, 18–20

equivalent dose a measure of radiation dose to biological tissue 34

erosion the wearing away of rock or other surface matter such as soil 6–7, 9

extended objects astronomical objects that are not merely a point, such as the Moon or a galaxy 44

eyepiece the (smaller diameter) lens through which the observer looks when using a telescope 45

F

fibre a long thin thread or filament 13–14

field in physics, a space in which a particular force acts 6, 17, 30–32, 34

focal length the distance from the centre of the lens to the focal point 44

focal point rays of light that strike a convex lens, parallel to the principal axis, converge to a focus at the focal point 43, 44

fold mountain a mountain caused by folding of the Earth's crust when two tectonic plates push against one another 7

force the push or pull that acts between two objects 21, 23–27, 30–31–33, 38

free electrons the outer electrons of atoms of materials that are good conductors which are loosely held and can break free easily so they can move freely 27–29, 32

frequency the number of waves passing a set point, or emitted by a source, per second 8–10, 12–14, 34, 38

fuel rod long narrow tube in a nuclear reactor which contains nuclear fuel in pellet form 17

G

galaxy group of billions of stars 4–5, 9

gamma rays (γ) ionising high-energy electromagnetic radiation from radioactive substances, harmful to human health 10–11, 14, 33–36

generator equipment for producing electricity 17, 20, 30, 32, 37

geologist scientist who studies rocks and the changes in the Earth 6

giant star stars that are 10–100 times larger and brighter than the Sun 60

global warming gradual increase in the average temperature of Earth's surface 12, 14–15, 19

globular clusters groups of older stars that surround a galaxy 52

gradient the degree of slope of a line 21, 25–26, 28

gravitational potential energy the energy an object gains due to its height 25–26

greenhouse effect the trapping of infra-red radiation by the Earth's atmosphere 12, 15

greenhouse gas a gas such as carbon dioxide that reduces the amount of infrared radiation escaping from Earth into space, thereby contributing to global warming 12, 14–15, 19

H

half-life the time taken for half of the atoms in a radioactive element to decay 35–38

hazard something that is likely to cause harm, e.g. a radioactive substance 18, 34–36

hertz unit for measuring wave frequency; 1 hertz (Hz) = 1 wave per second 8

Hertzsprung–Russell diagram a chart that plots the luminosity of a star (vertical axis) against its surface temperature (horizontal axis) 60

high level waste for example, (radioactive) spent fuel rods, with a long half-life, which need to be disposed of carefully 37–38

Hubble constant the ratio of the recessional velocity of a distant galaxy to its distance from Earth; measured in km/s/Mpc 53

hydroelectric description of power station generating electricity from the energy of moving water 17–20

I

ideal gas a theoretical model of a gas whose molecules take up no space and do not interact with each other. A gas that is well above its boiling point and at low pressure is a good approximation of an ideal gas 57

induced a term used to mean 'created' 30–32

in parallel when components are connected across each other in a circuit 28–29, 32

Glossary

in series when components are connected end-to-end in a circuit 28–29, 32

instantaneous speed the speed at a particular moment in time 21–22

insulator a substance in which electric current cannot flow freely 27, 32

intensity a measure of the power of a beam of radiation 10–12, 14, 32, 34

intermediate level waste for example, (radioactive) chemical sludge and reactor components, with short or longer half-lives that have to be disposed of with care 37–38

inversely proportional when there is an increase in one variable and a proportionate decrease in another variable 8, 10, 29, 57

ion atom (or groups of atoms) with a positive or negative charge, caused by losing or gaining electrons 33–34, 38

ionisation the removal of electrons from atoms or molecules 10, 34, 38, 55

ionising radiation electromagnetic radiation that has sufficient energy to ionise the material it is absorbed by 10–11, 14, 18, 20, 33–34, 36, 38

irradiation exposure to waves of radiation 18, 20, 36, 38

isotopes atoms that have the same number of protons, but different numbers of neutrons. Different forms of the same element 36–37, 59

J

joule unit of energy 15, 25, 28, 37

K

kilowatt unit of power equal to 1000 watts or joules per second 15, 20

kilowatt-hour (kWh) the energy transferred in 1 hour by an appliance with a power rating of 1 kW (sometimes called a 'unit' of electricity) 15, 20

kinetic energy the energy an object has due to its motion 23, 25–26, 37

L

lava molten rock (magma) from beneath the Earth's surface when it erupts from a volcano 6

light dependent resistor (LDR) a semiconductor device, where resistance changes with the amount of light 29

light pollution excessive artificial light that prevents us from seeing the stars at night and can disrupt ecosystems 4

light-year the distance travelled by light in 1 year 4, 50

limiting friction the maximum amount of force that can be applied to an object before it will move 23

line spectrum a series of coloured lines emitted by a vapour. The line spectrum from each element is unique 55

longitudinal a wave such as a sound wave in which the disturbances are parallel to the direction of energy transfer 7, 12

low level waste for example, contaminated (radioactive) paper and clothing that is not very dangerous, with a short half-life, but still needs to be disposed of carefully 37–38

luminosity total power emitted by a star in all directions, across all wavelengths 50

lunar eclipse when the Moon passes into the Earth's shadow 40

lunar month the time between two successive new (or full) moons, equal to 29 days, 12 hours, 44 minutes, 2.8 seconds 40

M

magma molten (liquid) rock 7

magnetic field a space in which a magnetic material exerts a force 6, 17, 30–32

main sequence the stage in a star's life where the fusion of hydrogen into helium takes place in the core 60

mantle semi-liquid layer of the Earth beneath the crust 6–7, 9

mass number the total number of protons and neutrons in a nucleus 59

metal a group of materials (elements or mixtures of elements) with broadly similar properties, such as being hard and shiny, able to conduct heat and electricity, and able to form thin sheets (malleable) and wires (ductile) 11, 19, 27, 29, 32

methane a gas with molecules composed of carbon and hydrogen; a greenhouse gas 12, 14, 18

microwave electromagnetic wave similar to radio waves but with higher energy 11, 13–14, 16

Milky Way the galaxy in which our Sun is one of billions of stars 4, 9

molecule two or more atoms held together by strong chemical bonds 10–11, 27, 34

momentum the product of mass and velocity; momentum (kg m/s) = mass (kg) × velocity (m/s) 24, 26

moon a large natural satellite that orbits a planet 4, 9

motor an electric motor converts electrical energy into mechanical energy 28, 31–32

motor effect a term used to describe the force experienced when a current flows through a wire in a region where there is a magnetic field. If it is free to move this is known as the motor effect 31–32

mutation a change in the DNA in a cell 34

N

National Grid the network that distributes electricity from power stations across the country, using cables, transformers and pylons 18, 20

neutral in physics, an atom with no overall charge 27, 34

neutrino a neutral particle of very low mass 59

neutron small particle that does not have a charge – found in the nucleus of an atom 27, 33–35, 37–38

neutron stars very dense remnants of a giant star, following a supernova 61

noise random alteration to a communication signal, possibly due to interference 13, 19

normal a line drawn at right angles to an interface between two materials, such as air and glass; used to assist in drawing ray diagrams 43

nuclear fission a chain reaction employed in nuclear power reactors in which atoms are split, releasing huge amounts of energy 5, 37–38

nuclear fuel radioactive fuel, such as uranium or plutonium, used in nuclear power stations 15, 37–38

nuclear fusion nuclear reaction in which two small atomic nuclei combine to make a larger nucleus, with a large amount of energy released 5, 37–38, 58

nucleus the central core of an atom, which contains protons and neutrons and has a positive charge 27, 33–35, 37–38

O

objective lens the (larger diameter) lens at the front of a refracting telescope 45

oceanic ridge undersea mountain range formed by seafloor spreading and caused by the escape and solidification of magma where tectonic plates meet 6

Ohm's law law that states that the current through a metallic conductor is directly proportional to the voltage across its ends, if the conditions are constant 28, 32

optical fibre glass fibre that is used to transfer communication signals as light or infrared radiation 13, 17

orbit near-circular path of an astronomical body around a larger body 4, 6

orbits electrons are arranged in orbits (or shells) around the nucleus of an atom 27, 33, 38

Glossary

oscilloscope laboratory equipment for displaying waveforms 8

ozone gas found high in the atmosphere which absorbs ultraviolet rays from the Sun 11, 14

P

p–p cycle a sequence of nuclear fusion reactions that takes place in small to average sized stars 59

parallax angle between two imaginary lines from two different observation points on Earth to an object such as a star or planet, used to measure the distance to that object 4, 9, 49

parsec astronomical unit of distance, equal to 3.26 light years 49

period time taken for one complete cycle, for example, of brightness changes by a variable star 51

photon a 'packet' of electromagnetic energy, the amount of energy depending on the frequency of the electromagnetic wave 10, 14

photosphere the region of a star from which light is emitted 54, 61

pixel a tiny area (for example a dot or square) on a screen which conveys the data relating to a small part of a picture 13

planet large sphere of gas or rock orbiting a star 4, 9

planetary nebula a shell of gas ejected from a star towards the end of its life. Nothing to do with planets! 61

plasma a gas of electrons and ions 59

plastic a compound produced by polymerisation, capable of being moulded into various shapes or drawn into filaments and used as textile fibres 16, 27, 32

plate boundary where two adjacent tectonic plates of the Earth's crust meet or are moving apart 7

positron a positively charged particle. The anti-matter equivalent of the electron 59

potential difference (p.d.) another term for voltage, a measure of the energy carried by the electrical charge 28–29, 32

power amount of energy that something transfers each second, measured in watts (or joules per second) 15–16, 18–20, 27–33

power (of a convex lens) a measure of how much the lens converges light 44

primary coil the input coil of a transformer 31–32

primary energy source a source of energy before conversion to useful energy; examples include fossil fuels, wind, biomass and solar energy 15, 17, 20

principal axis a line drawn through the centre of the lens and perpendicular to it; used to assist in drawing ray diagrams 44

principal frequency the main frequency of electromagnetic radiation emitted by an object; hotter objects have higher principal frequencies 12

proton small positively charged particle found in the nucleus of an atom 27, 33–35, 38

P-waves longitudinal shock waves following an earthquake that can travel through the molten core of the Earth; they change direction at the boundary between different layers of the Earth 7, 9

Q

quasar quasi-stellar object: distant object that emits immense power from a relatively small region of space 48

R

radiation energy transfer by electromagnetic waves or fast-moving particles 4–5, 10–12, 14, 18, 20, 33–36, 38

radiation zone the region of a star where most of the energy is transported by photons 54

radioactive a material that randomly emits ionising radiation from its atomic nuclei 11, 18–20, 33–38

radioactive decay the disintegration of a radioactive substance, the process by which an atomic nucleus loses energy 35

radioactive tracer a radioactive isotope with a short half–life that can be ingested and traced through a patient's body or to monitor the movement of waste products in industry 36

radiographer medical worker who takes and processes body images 11, 36

radiotherapy a technique that uses ionising radiation to kill cancer cells in the body 36

rate a measure of speed; the number of times something happens in a set amount of time 15, 17, 20, 22, 24, 27–28, 30, 33, 35

rate of flow of charge the current, or charge, that flows per second, measured in Ampères 27

rating an assessment or classification according to a scale, as in electrical appliances that are rated in terms of power or energy efficiency 15

ray diagram drawn to show the path of light through an optical system, such as a telescope; often used to show image formation by a lens 43

reaction force an equal force that acts in the opposite direction to the action force 23, 26

reactor the part of a nuclear power station where energy is released from nuclear fuel 17, 37

real brightness a measure of the light emitted by a star compared to the Sun, taking into account how far away it is 4

redshift the shift of lines in a spectrum towards the red (longer wavelength) end, due to the motion of the source away from us 5, 9, 48, 53

reflect in the case of light, re-direction of the light wave, usually back to the point of origin from a shiny surface 10–12

refraction the change in direction of a wave as it travels from one medium to another, due to the change in wave speed 43

relative brightness the apparent brightness of a star as seen from Earth; a dim star close to Earth may appear brighter than a bright one that is further away 4

repulsive a force that pushes two objects apart 23, 33

resistance a measure of how hard it is for an electric current to flow through a material 24–25, 28–29, 32

resolution ability to distinguish detail in an image or to recognise two nearby objects as distinct 47

resultant (force) the overall forces acting on an object added together 23–24, 26

retrograde motion the backward motion of a planet over a number of days, as seen against the background of stars 41

right ascension one of two angles that describe a star's location on the celestial sphere. Right ascension is similar to longitude and gives the star's position relative to a point on the celestial equator. Usually measured in units of time: hours, minutes and seconds 42

risk the likelihood of a hazard causing harm 11, 14, 18–19, 34, 36–38

S

Sankey diagram diagram showing how the energy supplied to something is transferred into 'useful' or 'wasted' energy 16, 18

seafloor spreading an extension of the seafloor caused by tectonic plate movement and the extrusion of magma between two plates which solidifies to form rock 6

secondary coil the output coil of a transformer 31, 33

Glossary

secondary energy source more convenient form of energy, such as electricity and refined fuels, produced from primary energy sources 15, 20

sediment particles of rock etc. in water that settle to the bottom 6, 8

sedimentary rock rock formed when sediments are laid down and compacted together 6

sedimentation the settling of particles in water to the bottom 6

seismic waves vibrations that pass through the Earth following an earthquake 7

sidereal day the time taken for a specific star to be in the same position in the sky on two successive nights 39

sievert (Sv) the SI unit of equivalent dose 34

signal information that is transmitted by, for example, an electrical current or an electromagnetic wave 13

solar day the time taken for the Sun to reach the highest point in the sky on two successive occasions 39

solar eclipse when the Moon passes between the Sun and the Earth, casting a shadow onto the Earth and obscuring the Sun 40

solar system the planetary system around the Sun, of which the Earth is part 3

speed how fast an object travels, calculated using the equation: speed (metres per second) = distance/time 21

speed–time graph a visual way of showing how an object's speed changes over a period of time 22

step down transformer device used to change the voltage of an a.c. supply to a lower voltage 31

step up transformer device used to change the voltage of an a.c. supply to a higher voltage 31

sterile containing no living organisms 34

strong nuclear force a force that holds all the particles together in a nucleus of an atom 33

supergiants massive stars that can fuse heavier elements and emit enormous power. When they run out of fuel they explode in a supernova, eventually forming a black hole 60

supernova explosion of a large star at the end of its life 5, 61

S-waves transverse shock waves following an earthquake that cannot travel through the molten core of the Earth 7

T

tectonic plate section of Earth's crust that slowly moves relative to other plates 7

terminal velocity the maximum speed achieved by any object falling through a gas or liquid 24

theory a creative idea that may explain an observation and that can be tested by experimentation 5, 6

thermistor a semiconductor device in which resistance changes with temperature 29

transect a line of quadrats 34

transmitted radiation that passes through a material 34

transverse a wave in which the disturbances are at right angles to the direction of energy transfer 7

turbine device which makes a generator spin to generate electricity 17

U

Universe the whole of space and all the objects and energy within it 4–5

V

vacuum a space where there are no particles of any kind 4, 10, 37

variable resistor a device that allows the amount of resistance in a circuit to be varied 28

velocity the speed of an object in a certain direction 21

velocity–time graph a visual way of showing direction of travel and acceleration 22

volcano landform from which molten rock erupts onto the surface 6–7

volt the unit of voltage 15

voltage a measure of the energy carried by an electric current (see potential difference) 28, 30–31

voltmeter a device used to measure the voltage across a component 28

W

watt (W) unit of power, or rate of transfer of energy, equal to a joule per second 15

wave a periodic disturbance that transfers energy 7–8, 10–11, 13

wave equation the speed of a wave is equal to its frequency multiplied by its wavelength 8

wave technology in renewable energy, equipment that allows us to harness the power of ocean waves 18

wavelength distance between two successive wave peaks (or troughs, or any other point of equal disturbance) 5, 8–9

white dwarf a small dense star of high surface temperature where fusion has stopped. The remnant of an average star at the end of its life 60

wind turbine device that uses the energy in moving air to turn an electricity generator 18, 20

work work is done when a force moves an object 25–26

X

X-rays ionising electromagnetic radiation 10–11, 14

Data sheet

Fundamental physical quantity	Unit(s)
length	metre (m); kilometre (km); centimetre (cm); millimetre (mm); nanometre (nm)
mass	kilogram (kg); gram (g); milligram (mg)
time	second (s); millisecond (ms); year (a); million years (Ma); billion years (Ga)
temperature	degree Celsius (°C); kelvin (K)
current	Ampère (A); milliAmpère (mA)

Prefixes for units			
nano (n)	one thousand millionth	0.000 000 001	$\times 10^{9}$
micro (μ)	one millionth	0.000 001	$\times 10^{-6}$
milli (m)	one thousandth	0.001	$\times 10^{-3}$
kilo (k)	$\times$ one thousand	1 000	$\times 10^{3}$
mega (M)	$\times$ one million	1 000 000	$\times 10^{6}$
giga (G)	$\times$ one thousand million	1 000 000 000	$\times 10^{9}$
tera (T)	$\times$ one million million	1 000 000 000 000	$\times 10^{12}$

Useful equations
speed = distance travelled ÷ time taken
acceleration = change in velocity ÷ time taken
momentum = mass × velocity
change of momentum = resultant force × time it acts
change in gravitational potential energy = weight × height difference
kinetic energy = ½ × mass × [velocity]^2
resistance = voltage × current
energy transferred = power × time
electrical power = voltage × current
efficiency = (energy usefully transferred ÷ total energy supplied) × 100%

Exam tips

The key to successful revision is finding the method that suits you best. There is no right or wrong way to do it.

Before you begin, it is important to plan your revision carefully. If you have allocated enough time in advance, you can walk into the exam with confidence, knowing that you are fully prepared.

Start well before the date of the exam, not the day before!

It is worth preparing a revision timetable and trying to stick to it. Use it during the lead up to the exams and between each exam. Make sure you plan some time off too.

Different people revise in different ways, and you will soon discover what works best for you.

Remember!

There is a difference between *learning* and *revising*.

When you revise, you are looking again at something you have already learned. Revising is a process that helps you to remember this information more clearly.

Learning is about finding out and understanding new information.

Some general points to think about when revising

- Find a quiet and comfortable space at home where you won't be disturbed. You will find you achieve more if the room is ventilated and has plenty of light.

- Take regular breaks. Some evidence suggests that revision is most effective when tackled in 30 to 40 minute slots. If you get bogged down at any point, take a break and go back to it later when you are feeling fresh. Try not to revise when you're feeling tired. If you do feel tired, take a break.

- Use your school notes, textbook and this Revision guide.

- Spend some time working through past papers to familiarise yourself with the exam format.

- Produce your own summaries of each module and then look at the summaries in this Revision guide at the end of each module.

- Draw mind maps covering the key information on each topic or module.

- Review the **Grade booster checklists** on page 148–156.

- Set up revision cards containing condensed versions of your notes.

- Prioritise your revision of topics. You may want to leave more time to revise the topics you find most difficult.

Workbook

The **Workbook** (pages 85–147) allows you to work at your own pace on some typical exam-style questions. These are graded to show the level you are working to (G–E, D–C or B–A*). You will find that the actual GCSE questions are more likely to test knowledge and understanding across topics. However, the aim of the Revision guide and Workbook is to guide you through each topic so that you can identify your areas of strength and weakness.

The Workbook also contains example questions that require longer answers (**Extended response questions**). You will find one question that is similar to these in each section of your written exam papers. The quality of your written communication will be assessed when you answer these questions in the exam, so practise writing longer answers, using sentences. The **Answers** to all the questions in the Workbook can be cut out for flexible practice and can be found on pages 161–168.

Collins Workbook

NEW GCSE

Physics

OCR

Twenty First
Century Science

Authors: Michael Brimicombe
Nathan Goodman
Sarah Mansel

Revision Guide +
Exam Practice Workbook

Our solar system and the stars

G–E

1 Our solar system contains asteroids. Name five other types of object in the solar system.

.... [5 marks]

D–C

2 Here are four planets in the solar system: Earth, Jupiter, Mercury, Uranus.

  a Put these planets in order of increasing mass.

.... [3 marks]

  b Put these planets in order of increasing size of orbit.

.... [3 marks]

B–A*

3 Explain why a light-year is about 1×10^{13} km.

....

....

.... [3 marks]

G–E

4 Complete the sentences by choosing words from this list.

galaxies   stars   planets   moons

The Milky Way is one of many .... in the Universe. Each galaxy is made of

many .... [2 marks]

D–C

5 A scientist has a hypothesis that the Sun is just another star. Suggest how they could attempt to disprove this hypothesis.

....

....

.... [3 marks]

B–A*

6 Explain why observing very distant galaxies is equivalent to observing the Universe as it was a long time ago.

....

....

....

.... [3 marks]

G–E

7 Here are some objects in the Universe: Earth, Milky Way, solar system, Sun. Put them in order of size, starting with the smallest.

.... [3 marks]

G–E

8 Explain how the brightness of a star can be used to estimate its distance.

....

....

....

.... [3 marks]

B–A*

9 Explain why the distance to most stars is not accurately known.

....

....

.... [2 marks]

The fate of the stars

1 Where does the Sun get its energy from?

.... [1 mark]

G–E

2 Shortly after the Big Bang, the only elements in the Universe were hydrogen and helium. Now the Universe contains many different elements. Explain where they came from.

....

....

....

....

G–E

.... [4 marks]

3 Explain why our solar system with solid planets must be much younger than the Universe.

....

....

B–A*

.... [2 marks]

4 Explain what we know about the motion of galaxies.

....

....

....

....

D–C

.... [4 marks]

5 Draw lines to link the start of each sentence to its correct ending.

The Sun is ...		... 14 000 million years old.
The Earth is ...		... 5000 million years old.
The Universe is ...		... 4500 million years old.

G–E

[2 marks]

6 Scientists believe that the Universe around us today came into being 14 000 million years ago. Explain why they are much less certain about what will happen to the Universe in the future.

....

....

....

....

....

B–A*

.... [4 marks]

Earth's changing surface

1 Draw lines to link the start of each sentence to its correct ending.

Fossils are ...		... new mountains made from lava.
Mountains are ...		... materials from the erosion of mountains.
Volcanoes are ...		... the remains of plants and animals in rock.
Sediments are ...		... made by folding the rocks of the Earth's crust.

[3 marks]

2 Describe and explain the formation of sedimentary rocks.

....

....

....

....

[4 marks]

3 The Earth is about 4000 million years old. Explain why erosion has not reduced all the continents to sea level in that time.

....

....

....

[4 marks]

4 Which of these observations provide evidence for Wegener's theory of continental drift? Put a tick (✓) next to your answer.

    a There is life on all of the continents. ☐

    b The continents fit together like a jigsaw. ☐

    c Many continents are surrounded by the sea. ☐

    d There are sedimentary rocks on all of the continents. ☐

    e There are chains of mountains at the edges of some continents. ☐ **[2 marks]**

5 Give reasons why scientists didn't accept Wegener's theory of continental drift when it was first published.

....

....

....

....

....

[5 marks]

6 Describe and explain the pattern of magnetisation of rocks on the Atlantic seafloor.

....

....

....

[4 marks]

Tectonic plates and seismic waves

1 State **three** things which happen at plate boundaries.

....

.... [3 marks] **D–C**

2 What is a plate boundary?

.... [2 marks]

3 Explain the presence of volcanoes at plate boundaries.

....

....

....

.... [4 marks]

4 Explain why earthquakes happen at plate boundaries.

....

.... [1 mark] **B–A***

5 Although the Earth was formed about 4000 million years ago, the majority of the rocks on its surface are much younger. Use the rock cycle to explain why.

....

....

....

.... [4 marks]

6 Draw lines to link each part of the Earth with its best description.

core	solid rock
crust	liquid iron
mantle	semi-liquid rock

[2 marks] **G–E**

7 Complete the sentences by choosing the correct words from the list below.

| continent | faster | slower | P-waves |
| S-waves | seismic | sound | tectonic |

Sudden movement of .... plates makes two types of .... waves.

The ones which can only move through the crust and mantle are called ....

The ones which can also move through the core are called ....

P-waves move .... than S-waves. [5 marks] **D–C**

8 Explain the evidence from earthquakes of the structure of the Earth.

....

....

....

.... [4 marks] **B–A***

Waves and their properties

1 Draw lines to link each property of a wave to its correct description.

| Speed | | Maximum value of the disturbance in one wave |

| Frequency | | How far the energy of the wave travels in a second |

| Amplitude | | The number of vibrations of the wave source in one second |

| Wavelength | | Distance along wave from one zero disturbance to the next |

[3 marks]

2 What does a wave carry away from its source?

.... **[1 mark]**

3 Ben watches a firework display. He hears the bang from a firework 1.5 seconds after he sees the flash. How far away was the firework?

(Speed of sound = 300 m/s, speed of light = 300 000 000 m/s)

....

....

.... **[2 marks]**

4 Draw lines to link each quantity with its correct units.

| Speed | | m |

| Frequency | | Hz |

| Wavelength | | m/s |

[2 marks]

5 Complete the sentence by choosing the correct word from the list below.

height   length   number   time

The frequency of a wave is the .... of vibrations per second produced by it.   **[1 mark]**

6 What is the speed of a wave which has a frequency of 20 kHz and a wavelength of 40 cm?

....

.... **[2 marks]**

7 a The speed of red light in glass is 2.0×10^8 m/s. The wavelength of red light in glass is 0.44 μm. Calculate its frequency.

....

.... **[2 marks]**

b Blue light has a frequency of 6.7×10^{14} Hz. Estimate its wavelength in glass.

....

....

.... **[3 marks]**

P1 Extended response question

Explain the evidence from distant galaxies for an expanding Universe.

✏ *The quality of written communication will be assessed in your answer to this question.*

B–A*

[6 marks]

Waves which ionise

1 Sentences a–d explain how you see this page. Put them in the correct order (1–4).

a The light travels away from the source. ☐

b A light source in the room emits some light. ☐

c The light is absorbed by detectors in your eyes. ☐

d The light bounces off the paper but is absorbed by the ink. ☐ **[3 marks]**

2 How fast does light pass through a vacuum? .... **[1 mark]**

3 Arrange these parts of the electromagnetic spectrum in order of increasing energy: infrared;

gamma rays; radio waves; ultraviolet. .... **[3 marks]**

4 Complete these sentences about the transfer of energy by X-rays. Choose words from this list.

amplitude   energy   frequency   light   photons

power   protons   speed   sound

X-rays transfer .... in packets called .... from one place to another

at the speed of .... Increasing the energy of a photon requires an increase of

its .... **[4 marks]**

5 Tick (✓) to show which of these changes increases the intensity of an electromagnetic wave.

a Increase the speed of the photons ☐   **c** Increase the energy of each photon ☐

b Increase the frequency of the wave ☐   **d** Increase the number of photons per second ☐ **[3 marks]**

6 Explain why the intensity of a wave decreases as it moves away from its source.

....

.... **[3 marks]**

7 A laser beam transfers 4.8 mJ of energy in 8.0 s to a black surface of area 2.0×10^{-6} m^2.
Calculate the intensity of the laser beam.

.... **[2 marks]**

8 Explain what happens to an atom when one of its electrons is knocked out.

....

.... **[3 marks]**

9 a Which parts of the electromagnetic spectrum are ionising radiations?

.... **[3 marks]**

b Explain how an electromagnetic wave can ionise an atom.

....

.... **[2 marks]**

10 Explain why exposure to ionising radiation is harmful to people.

....

....

....

.... **[5 marks]**

Radiation and life

1 Here are four statements about ionising radiation.

   i They are absorbed by muscle but not by bone.   iii They pass easily through both muscle and bone.

   ii They are absorbed by bone but not by muscle.   iv They are produced by radioactive materials.

   a Which of the statements are **only** true for X-rays? .... [1 mark]

   b Which of the statements are **only** true for gamma rays? .... [2 marks]

2 Complete the sentences. Choose words from this list.

   absorbed   clothes   goggles   increasing   reducing

   reflected   shields   stopping   transmitted

   X-rays can be used to image people's luggage at airports. The X-rays are .... by

   high density items but are .... by the low density items. The people who operate

   these X-ray scanners can reduce their risk of cancer with .... made of lead

   and .... the time they spend near the scanner.   [4 marks]

3 Which of these actions **increase** the amount of thermal energy transferred from a light source to an object?

   a Leave the light switched on for longer   ☐

   b Increase the intensity of the light source   ☐

   c Increase the frequency of the light radiation   ☐

   d Put the light source and object in a darkened room   ☐

   e Increase the distance of the object from the light source   ☐   [3 marks]

4 Explain why dry food does not heat up in a microwave oven.

   ....

   .... [2 marks]

5 Describe how scientists study people to determine the risk of using mobile phones.

   ....

   ....

   ....

   .... [5 marks]

6 a Name the electromagnetic radiation in sunlight which gives you sunburn. .... [1 mark]

   b In what other way can this radiation damage you? .... [1 mark]

   c What can you do to protect yourself when exposed to sunlight?

   .... [2 marks]

7 Sunbathing has both risks and benefits.

   a What are the risks? .... [2 marks]

   b What are the benefits? .... [2 marks]

8 What is the effect of ultraviolet radiation on the ozone layer?

   ....

   .... [2 marks]

Climate and carbon control

1 Tick (✓) to show which of these statements about radiation from the Sun is correct.

G-E

a It contains all possible frequencies ☐   **d** All of it passes through our atmosphere ☐

b It contains a range of frequencies ☐   **e** It contains just one frequency – light ☐

c All of it is absorbed by our atmosphere ☐   **f** Some is absorbed by our atmosphere ☐   **[2 marks]**

2 Complete the sentences. Choose the best words from this list.

all	average	cold	decreasing	enduring

greatest	increasing	hot	least

Electromagnetic radiation is emitted by .... objects. The principal frequency

of that radiation is the one with the .... intensity, and it increases

with .... temperature.   **[3 marks]**

3 The diagram represents the flow of some carbon on the Earth.

Match each arrow with what it represents.

i Animals eat plants for food ☐

ii Animals breathe out carbon dioxide ☐

iii Plants use carbon dioxide to build tissues ☐

iv Plants emit carbon dioxide from respiration ☐

[3 marks]

4 Explain why the level of carbon dioxide in the atmosphere has been steady for thousands of years.

....

....   **[3 marks]**

5 The temperature of both the atmosphere and the amount of carbon dioxide in the atmosphere has been rising for the past century. How do scientists determine which of these factors is the cause and which is the effect?

....

....

....   **[3 marks]**

6 State three effects of global warming which could affect food grown in the UK.

....

....   **[3 marks]**

7 Explain how scientists can test and use a climate model to predict the effects of global warming.

....

....

....

....   **[4 marks]**

Digital communication

1 The waves of the electromagnetic spectrum are: gamma rays; infrared; microwaves; radio; visible; ultraviolet; X-rays. Which four do we use for communication?

.... [4 marks]

G–E

2 Explain why radio waves are used to carry TV broadcasts.

.... [2 marks]

D–C

3 Explain why infrared is used for long-distance telephone communication.

.... [2 marks]

4 What effect does modulation have on a carrier wave?

....

B–A*

.... [2 marks]

5 Which one of these graphs shows a digital signal? .... [1 mark]

signal   signal   signal

      time   time   time

      A   B   C

G–E

6 Sound information is carried by a radio wave as a code using 1 and 0.

a What are 1 and 0 in this context?

.... [2 marks]

b Explain how the radio receiver is able to reproduce the sound information.

....

....

D–C

.... [4 marks]

7 Explain why digital radio receivers are often unaffected by background radio noise.

....

B–A*

....

.... [3 marks]

8 Complete the passage about digital information. Choose from this list: *0, 1, 2, 4, 8.*

Digital information for sound or pictures has only .... values. Sound information is coded

as a string of binary digits called .... or .... and stored in groups of ....

as bytes.

G–E

[4 marks]

9 Explain how sounds can be created by an MP3 player from digital information.

....

....

D–C

.... [4 marks]

10 List four advantages of using digital signals to transfer information from one place to another.

....

B–A*

.... [4 marks]

Explain how cutting down forests affects the amount of carbon dioxide in the atmosphere.

✎ *The quality of written communication will be assessed in your answer to this question.*

D–C

[6 marks]

Energy sources and power

1 State the names of **three** different fossil fuels.

.... [3 marks]

2 State the names of **four** different renewable energy sources.

....

.... [4 marks]

3 Why is electricity called a secondary energy source?

....

[1 mark]

4 Explain why the way that we make our electricity will have to change in the future.

....

.... [2 marks]

5 Explain the environmental impact of building more gas-fired power stations in the UK.

....

....

.... [4 marks]

6 Complete the sentences. Choose words from this list.

energy   force   joules   newtons   seconds   watts   weight

The power of a component is the .... which transfers to it in one second.

It is measured in .... when the energy is in .... and the time

in .... [4 marks]

7 A 250 W computer is left on for 24 hours. How much energy in kWh does it consume?

.... [2 marks]

8 A kettle is plugged into the mains electricity supply. Link the start of each sentence about the kettle to its correct ending.

The power of the kettle ...	... is energy transfer per second.
The energy in the circuit ...	... provides the current with energy.
The current in the circuit ...	... transfers energy from the supply.
The voltage of the supply ...	... flows from the supply to the kettle.

[3 marks]

9 Complete the equation for a microwave oven.

The .... of the oven (in watts) = the supply voltage (in ....) ×

the .... (in ....) [4 marks]

10 An appliance draws a current of 3.0 A from the 230 V mains supply. At what rate is energy transferred by the electric current?

.... [2 marks]

11 What is the current drawn from the 230 V mains supply when an appliance of power 1.15 kW is plugged in?

....

.... [2 marks]

G–E

D–C

B–A*

G–E

D–C

B–A*

Efficient electricity

1 Draw lines to link the start of each sentence about measuring electricity to its correct ending.

One kilowatt-hour is ...	... a decimal code.
	... 3 600 000 joules.
	... units of kilowatt-hours.
The meter readings are in ...	... a unit of electrical power.
	... the energy transferred into a house.
	... the current transferred out of a house.
An electricity meter records ...	... the voltage of the mains supply of a house.

[3 marks]

2 A washing machine has an average power rating of 800 W when it is run off a 120 V mains supply. A single wash cycle takes 90 minutes. If a unit of electricity costs 15 p, how much does the electricity for a wash cycle cost?

....

.... **[3 marks]**

3 Here is the Sankey diagram for a TV set.

   a What does the Sankey diagram show?

   ....

   ....

14 J sound

42 J input

18 J heat   **[2 marks]**

   b One of the labels of the diagram is missing. What should it be? .... **[2 marks]**

4 Link the start of each sentence about a Sankey diagram to its correct end.

An arrow splits ...	... out to the right.
Electrical energy flows ...	... in from the left.
Waste heat energy flows ...	... out downwards.
The electrical energy input is ...	... proportional to its energy.
The thickness of each arrow is ...	... where there is an energy transfer.
Useful transferred energy flows ...	... equal to the sum of the energy outputs.

[5 marks]

5 An electric drill transfers every 50 J of electricity into 20 J of useful kinetic energy. The rest is wasted as heat and sound. Calculate the percentage efficiency of the drill.

.... **[2 marks]**

6 Suggest three ways in which the government could make all of us use less energy.

....

.... **[3 marks]**

7 Explain why global demand for energy is likely to rise in the future.

....

....

....

....

[4 marks]

Generating electricity

1 Complete the sentences. Choose words from this list.

coil   generators   length   magnet   motors   transformers

Power stations use .... to make electricity. Each one contains a ....

which spins around near a .... of wire.   [3 marks] **G–E**

2 A magnet spins near a coil of wire. What appears across the wire? Circle the correct answer.

charge   electricity   power   voltage   [1 mark]

3 Sentences a–e describe how a power station makes electricity. Number the boxes and put them in the correct order (1–5).

a Water boils into steam. ☐

b A magnet is made to spin round. ☐

c A voltage appears across a coil of wire. ☐

d The high pressure gas spins the turbine round. ☐

e A primary fuel transfers energy to thermal energy. ☐   [4 marks] **D–C**

4 a State **three** primary energy sources which boil water into steam in a power station.

....   [3 marks]

b State **three** primary energy sources which spin turbines directly.

....   [3 marks]

5 What is the job of a turbine in a power station?

....

....   [2 marks] **G–E**

6 Many power stations use steam to spin the turbines. State **three** other substances used to spin turbines for electricity.

....   [3 marks]

7 Describe how fossil fuels are used to spin turbines in power stations.

....

....

....

....

....   [4 marks] **D–C**

8 Explain why a gas-fired power station has two different types of turbine.

....

....

....

....

....   [4 marks]

9 Describe in detail how the fuel transfers energy to a turbine in a nuclear power station.

....

....

....

....

....

....   [5 marks] **B–A***

Electricity matters

G–E

1 Why is nuclear waste a health risk?

.... [1 mark]

2 People who work with nuclear waste can be affected when they are irradiated by it.

  a Why is irradiation a risk to health?

.... [2 marks]

D–C

  b Suggest what steps people should take to reduce the risk of irradiation from nuclear waste.

.... [2 marks]

  c Explain why people who work with nuclear waste need to worry more about contamination.

....

....

....

.... [3 marks]

B–A*

3 Explain how governments assess the risks of different technologies for making electricity.

....

....

.... [3 marks]

G–E

4 Here are some energy sources which can be used to spin turbines directly. Circle the three renewable ones.

gas　　　　hydroelectric　　　　tidal　　　　wind　　　　　　　　　　 [2 marks]

D–C

5 Hydroelectricity is renewable. Explain other advantages and disadvantages of using hydroelectricity.

....

....

....

.... [4 marks]

B–A*

6 Explain how the use of renewable technology can reduce the environmental impact of generating electricity.

....

....

....

.... [4 marks]

D–C

7 What is the voltage of the mains supply in our homes?

.... [1 mark]

B–A*

8 The efficiency of transfer of energy from coal in a power station to the mains supply in your house is less than 50%. State the sources of wasteful energy transfer in order of their importance.

....

....

.... [4 marks]

Electricity choices

1 Draw lines to link each energy source with its impact on the environment.

Energy source		Impact on environment
Fossil fuel		Noise and visual pollution
Wind power		Floods large areas of land
Nuclear power		Produces radioactive waste
Hydroelectricity		Produces greenhouse gases

[3 marks]

2 Name four different energy sources which will run out in the next thousand years.

.... [4 marks]

3 Which of these energy sources do not contribute to global warming?
Circle your answers.

coal   gas   geothermal   nuclear   oil   solar   wind [4 marks]

4 The government decides to increase generating capacity in the UK by 20 000 MW. They shortlist these two ways of doing this: 1000 MW nuclear power stations; 500 MW wind farms.

Discuss what they should decide to do. Justify your recommendation.

....

....

....

....

....

[4 marks]

5 The USA produces about twice as much carbon dioxide per person as the UK. Why should this be of concern to people in both countries?

....

.... [1 mark]

6 Explain why people in the UK will have to use less energy in the future to keep global energy demand unchanged.

....

....

....

....

[4 marks]

7 In planning for the future, governments are always anxious to avoid the possibility of power cuts. Explain what they could do to avoid power cuts.

....

....

....

....

.... [4 marks]

P3 Extended response question

Some power stations in the UK are coming to the end of their working life. The government has to decide on replacements for them. One factor which needs to be considered is the cost of building the replacement. What **other** factors should the government consider?

🖉 The quality of written communication will be assessed in your answer to this question.

D–C

[6 marks]

Speed

1 A cyclist travels 60 km in 3 hours. Calculate his speed in:

  a kilometres per hour: .... [2 marks]

  b metres per second: .... [2 marks]

2 How long would it take the cyclist in Question 1 to travel 100 km if he travels at the same average speed?

.... [2 marks]

3 The average speed of a train travelling from London to Birmingham is 150 km/h. Explain whether the train's average velocity will be lower or higher than its average speed.

....

....

.... [3 marks]

4 Match these descriptions of motion with the distance–time graphs below, by writing the correct description next to each graph:

Constant high speed   **Constant slow speed**   **Speeding up**   **Stationary**

a)   b)   c)   d)

[4 marks]

5 Jennifer draws a distance–time graph for a walk to the local shop and back again (shown opposite).

  a How far does Jennifer travel in the first 10 minutes?

.... [1 mark]

  b Describe the rest of Jennifer's journey.

....

....

....

.... [4 marks]

  c Calculate Jennifer's speed in metres per second during the first 10 minutes.

.... [2 marks]

  d What was the total distance for the whole journey? .... [1 mark]

  e What was the total displacement for the whole journey? .... [1 mark]

  f Calculate both the average speed and the average velocity for the whole journey, and comment on your answers.

....

....

....

.... [4 marks]

Acceleration

G-E

1 Car A does 0 to 100 km/h in 12 seconds. Car B does 0 to 100 km/h in 10 seconds.

Which of these cars has the greatest acceleration? .... **[1 mark]**

D-C

2 Convert 100 km/h into metres per second and calculate the acceleration of the two cars above in metres per second squared.

....

....

....

.... **[5 marks]**

B-A*

3 Joshi drops a ball out the window. It accelerates from rest at 10 m/s² for 1.6 seconds, then hits the ground. At what speed does the ball hit the ground?

....

.... **[2 marks]**

G-E

4 Diana cycles along a straight road. The graph shows how her speed changes during the first minute of her journey.

Describe how Diana's speed changes in as much detail as possible.

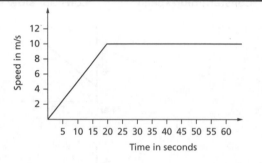

....

....

....

.... **[4 marks]**

5 The next day Diana cycles along the same stretch of road, but accelerates steadily for the first 20 seconds and then stays at a steady speed of 8 m/s for 20 seconds. Then she slows down steadily for 10 seconds until her speed is 5 m/s. Diana cycles at this slower speed for 10 seconds.

Draw a line on the graph in Question 4 to represent this journey. **[4 marks]**

6 The table below shows how the speed of a vehicle changes over a minute.

Time (s)	Speed (m/s)
0	0
5	3
10	6
15	9
20	12
25	14
30	16
35	16
40	16
45	16
50	12
55	8
60	4

D-C

a Plot a speed–time graph on the axes. **[4 marks]**

b Use the graph to calculate the acceleration during the first 20 seconds.

.... **[2 marks]**

B-A*

c Calculate the deceleration in the last 15 seconds.

....

.... **[4 marks]**

Forces

1 Complete the passage below using words from this list:

a downwards   **a sideways**   **an upwards**   **space**   **pairs**

Forces always occur in .... When you lean on a table,

you exert .... force from your hand onto the table.

The table exerts .... force on your hand.   [3 marks]

2 For each the following situations, one force is described. Describe the second force that makes up the interaction pair.

a A tennis racket pushes a tennis ball forwards.

....   [2 marks]

b A horse pulls a cart forwards.

....   [2 marks]

c The Moon is attracted towards Earth.

....   [2 marks]

3 Add to the diagram below to show the two forces acting on this book lying on top of a table. Label each force.

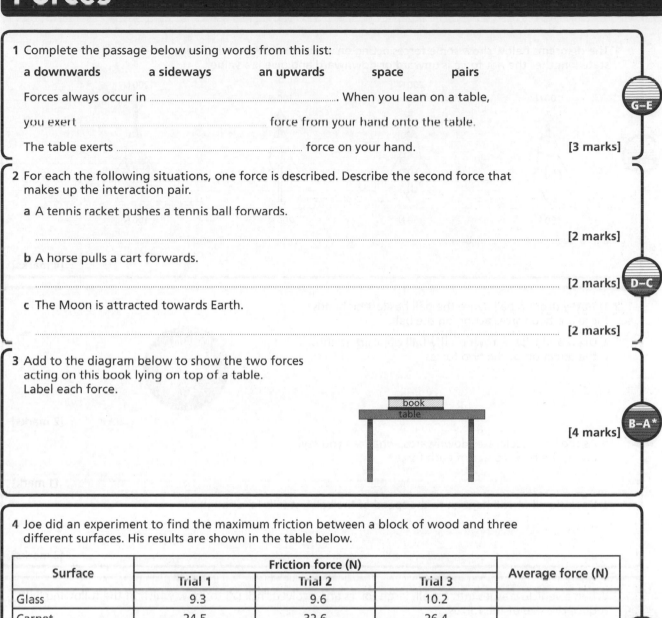

[4 marks]

4 Joe did an experiment to find the maximum friction between a block of wood and three different surfaces. His results are shown in the table below.

Surface	Friction force (N)			Average force (N)
	Trial 1	Trial 2	Trial 3	
Glass	9.3	9.6	10.2	
Carpet	24.5	32.6	26.4	
Wood	15.0	18.1	16.4	

a Why did Joe repeat the experiment three times for each surface?

....   [1 mark]

b Identify an anomalous reading. ....   [1 mark]

c What did Joe have to keep the same for all his trials in this experiment?

....   [1 mark]

d Calculate the average value for force for each surface. Add your answers to the table.

....

....   [3 marks]

e Which surface has the least friction? Explain your answer.

....

....   [2 marks]

5 Friction can be wanted or unwanted. On a bicycle, give one example of useful friction and one example of unwanted friction. Explain your answers.

....

....   [4 marks]

Effects of forces

1 The diagrams below show some forces acting on a spacecraft. For each diagram **a–d**, state whether the net force is **upward** or **downward** and give it a value.

G–E

100 N   200 N   100 N   200 N

150 N   150 N   200 N   100 N

a .... b .... c .... d ....

.... **[4 marks]**

2 Timothy drops a ball. Once the ball has left his hands there are two forces acting on the ball.

  a Draw and label arrows on the ball opposite to show the direction of the two forces.

[2 marks]

  b As the ball accelerates downwards, what can you say about the relative size of each force?

.... **[1 mark]**

  c Describe what will happen to the speed of the ball if it falls far enough.

....

....

[2 marks]

3 When a vehicle crashes into a wall, large forces are involved. Tick (✓) to show which of the following will reduce the size of the forces.

  a a larger (more massive) vehicle ☐   **c** increasing the time to stop ☐

  b slower speed ☐   **d** decreasing the time to stop ☐   **[2 marks]**

4 Calculate the momentum of the following objects:

  a A ball of mass 300 g travelling at a speed of 30 m/s.

.... **[2 marks]**

  b A car of mass 1100 kg travelling at a speed of 15 m/s.

.... **[2 marks]**

  c A lorry of mass 4000 kg travelling at a speed of 5 m/s.

.... **[2 marks]**

5 Class 11B are playing cricket. Nigel catches the ball when it is going fast and it makes his hands sting.

Using ideas about momentum, explain how Nigel could prevent his hands stinging.

....

....

....

....

.... **[4 marks]**

Work and energy

1 Complete the passage below, which explains a law about energy, using words from this list:

generated　　**conservation**　　**destroyed**　　**transferred**　　**devastation**

Energy can never be created or... It can only

be .... from one form to another or from one place to

another. This is called the .... of energy.　　**[3 marks]**

2 Tick (✓) to show which activities are examples of work being done.

　a lifting a weight ☐　　c kicking a football ☐

　b sitting on a chair ☐　　d watching a film ☐　　**[2 marks]**

3 Sophie lifted a ball weighing 3 N from the floor onto a shelf 1.3 m above the ground.

　a Calculate how much work Sophie did.

　...　**[2 marks]**

　b What sort of energy did the ball gain?

　...　**[1 mark]**

　c The ball then rolls off the shelf and falls to the floor. What is the maximum kinetic energy
　of the ball when it hits the floor?

　...　**[1 mark]**

　d What is the maximum speed at which the ball (of mass 0.3 kg) strikes the floor?

　...

　...　**[3 marks]**

4 A pendulum swings from side to side.

　a Mark on the diagram where the pendulum has maximum kinetic
　energy and where it has maximum potential energy.

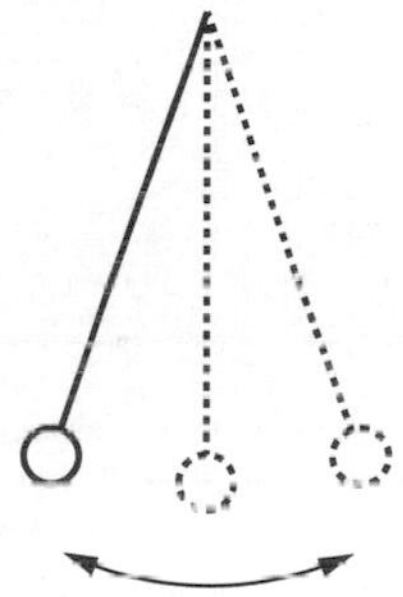

　[2 marks]

The pendulum bob has a weight of 1 N, and the height from which it
is released is 0.2 m above its lowest point.

　b Calculate the initial gravitational potential energy.

　...　**[2 marks]**

　c At one point the pendulum is travelling at a speed of 1.2 m/s and its mass is 0.1 kg.
　What is its kinetic energy?

　...　**[2 marks]**

　d At this point, what would be the gravitational potential energy of the pendulum bob?

　...　**[2 marks]**

5 a Describe the energy changes on a roller coaster.

　...

　...　**[3 marks]**

　b Explain why the second peak must be lower than
　the starting point of the roller coaster, as shown
　in the diagram.

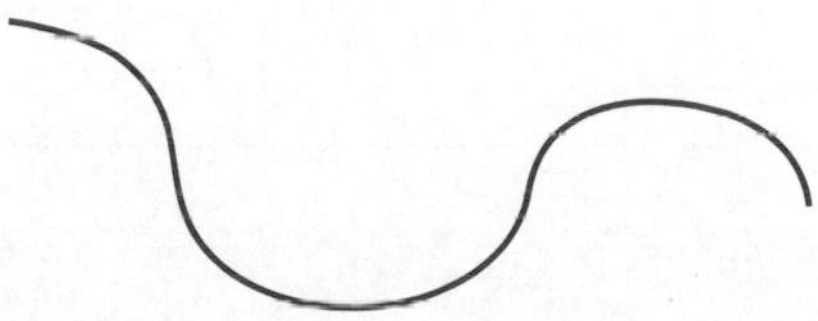

　...

　...

　...　**[3 marks]**

Describe the safety features that car manufacturers build into their cars.

🖉 *The quality of written communication will be assessed in your answer to this question.*

D–C

....

....

....

....

....

....

....

....

....

....

....

....

....

....

....

....

....

....

....

....

....

....

....

....

[6 marks]

Electric current – a flow of what?

1 Atoms contain electrons, neutrons and protons. Draw lines to connect the position in the atom with the particle and its charge.

Position	Particle	Charge
	Electron	Positive
Inside the nucleus		
	Proton	Neutral
Orbiting the nucleus		
	Neutron	Negative

[4 marks] G–E

2 Use words from this list to complete the paragraph below.

electrons   positive   negative   protons   neutrons   neutral   attracted   repelled

Lucy rubs a polythene rod with a duster. Some .... are transferred

from the duster to the rod. This makes the rod gain a ....

charge. When she puts the charged rod close to some small pieces of paper they are

.... to the rod.   [3 marks]

3 Explain why a metal rod will not become charged by rubbing.

....   [2 marks] D–C

4 Sadie is ironing some polyester shirts. When she has finished she switches off the iron at the wall socket and receives a small electric shock, which tickles her finger. Explain why.

....

....

....

                                                                                                            [5 marks] B–A*

5 John makes up a circuit to test some objects to find out whether they conduct electricity. He connects the objects between the crocodile clips in the circuit opposite.

Circle the objects in the list below that make the bulb light up.

2 p coin   plastic ruler   steel scissors   eraser   [2 marks]

G–E

6 Explain how energy is transferred from the cell to the light bulb in this circuit.

....

....

....   [5 marks] D–C

7 a In the circuit diagram for Question 6, draw an arrow to show the direction in which the charged particles flow when the current flows.   [1 mark]

b Explain why this happens.

....

....   [2 marks] B–A*

c Explain how you could make the bulb glow brighter.

....

....   [2 marks]

Current, voltage and resistance

1 John is trying to measure the current and voltage in a simple circuit. He needs to add a voltmeter and an ammeter to the circuit.

  a Add to the circuit diagram to show the correct symbols and positions for the voltmeter and ammeter.

[4 marks]

  b Describe what happens to the charged particles in the above circuit.

[3 marks]

2 What is meant by the term 'potential difference'?

[3 marks]

3 List the four circuits opposite in order of decreasing electrical current.

[3 marks]

4 A motor is connected to a power supply. When the voltage is 12 V the current is 0.5 A. Calculate the resistance (include the unit).

[3 marks]

5 John carried out an experiment to find the resistance of a piece of nichrome wire. He varied the potential difference across the piece of wire and recorded the current. The table shows his results.

Potential difference in volts	Current in amps
2	0.12
4	0.25
6	0.34
8	0.44
10	0.59
12	0.72

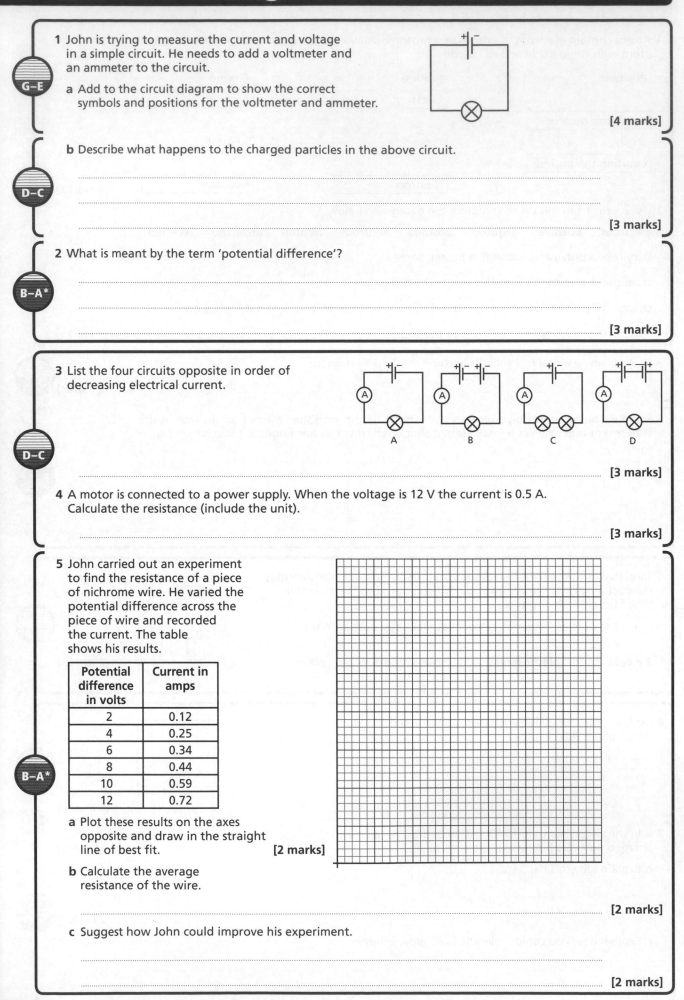

  a Plot these results on the axes opposite and draw in the straight line of best fit. **[2 marks]**

  b Calculate the average resistance of the wire.

[2 marks]

  c Suggest how John could improve his experiment.

[2 marks]

Useful circuits

1 Jody connects some bulbs in a circuit and measures the current. She first records the current with one cell and two bulbs in series.

   a What would happen to the current if Jody were to add another bulb in series?

   .... [1 mark]

   b What would happen to the current if Jody used a second cell in series?

   .... [1 mark]

   c What would happen to the current if one of the bulbs 'blew'?

   .... [1 mark]

G–E

2 Abdul is investigating some sets of Christmas tree lights. One set has 20 bulbs connected in parallel and the other has 20 similar bulbs connected in series. Both are connected to a 24 V power supply.

   a How could Abdul easily tell which was which?

   .... [1 mark]

   b What would happen to each set of lights if he unscrewed one bulb?

   ....

   .... [2 marks]

D–C

3 In the circuit shown below, what would be the readings on ammeters A_1, A_2, A_3 and A_4?

   A_1: ....

   A_2: ....

   A_3: ....

   A_4: ....

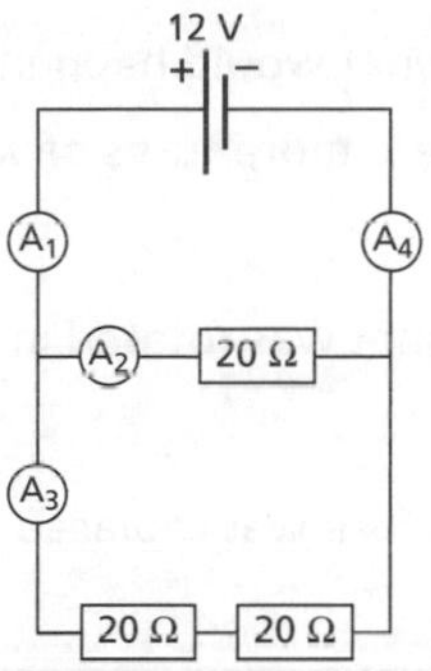

B–A*

[4 marks]

4 Describe the difference between an LDR and a thermistor.

....

.... [3 marks]

G–E

5 Thomas wants to use a thermistor as a thermometer. He needs to calibrate it before he can use it to measure temperatures. Describe how he could calibrate it.

....

....

.... [4 marks]

D–C

6 Rachel is investigating current in bulbs. She connects a bulb to a 12 V power supply and records the current flowing through the power supply as 0.7 A. She then connects a second identical bulb in series.

   a Why does Rachel expect the new current to be 0.35 A?

   ....

   .... [2 marks]

B–A*

   b The actual current is measured as 0.5 A. Explain why this current is higher than Rachel expected.

   ....

   ....

   .... [4 marks]

Producing electricity

1 Complete the paragraph below using words from this list:

sides **field** **metal** **magnetic** **iron** **wood** **compasses** **poles** **force**

A magnetic .... is the space around a magnet where

a .... material will experience a force. To see the shape of it we

can use .... filings or plotting ....

The force is stronger at the .... of the magnet. **[5 marks]**

2 The diagram shows a simple generator.

(*Source:* OCR (A182) *Twenty First Century Science – Physics A*)

Describe what would happen to the output if:

a there were more turns of wire in the coil.

.... **[1 mark]**

b the magnet was rotated in the opposite direction.

.... **[1 mark]**

c the iron core was replaced by an aluminium one.

.... **[1 mark]**

3 Some electrical devices use d.c. and some use a.c. Complete the tick chart below to show which devices use which type of electricity.

	a.c. (✔)	d.c. (✔)			a.c. (✔)	d.c. (✔)
a Torch			c Computer			
b Transformer			d Electric iron			

[4 marks]

4 The mains voltage in the UK is 230 V a.c. A bar heater draws a current of 3.5 A when one bar is switched on.

Calculate the power used by the heater.

....

.... **[3 marks]**

5 Julian decides to replace his filament bulbs with compact fluorescent tubes because they are more efficient.

a The power rating for the filament bulb is 100 W. Calculate the current flowing in the bulb when connected to the mains at 230 V.

....

.... **[3 marks]**

b If the compact fluorescent tube uses a quarter the current of the filament to produce the same amount of light, calculate the power rating for the compact fluorescent tube.

....

.... **[2 marks]**

Electric motors and transformers

1 Add to this diagram to show the magnetic field lines around the coil. The field lines will go from left to right.

D–C

[2 marks]

2 This motor is a simple coil suspended in a magnetic field. The coil starts in the position shown. When a p.d. is applied to the connecting wires the coil rotates.

coil

brushes

split-ring commutator

B–A*

Explain how the commutator and brushes enable the coil to experience continuous rotation.

....

....

....

....

[5 marks]

3 Complete the paragraph using words from this list to explain how transformers work.

primary	voltage	iron	secondary	direct
current	alternating	copper	coils	magnets

Transformers are used to change the .... of electrical power

supplies. They only work with .... current. A transformer consists

of two .... wrapped around an ....

core. The input voltage is connected across the .... coil and the

new voltage will be output across the .... coil.

G–E

[6 marks]

4 The diagrams below show four transformers (a–d), each with a different number of coils.

200 turns	10 turns	20 turns	1000 turns	20 turns	10 turns	200 turns	200 turns

230 V a.c. V_1 230 V a.c. V_2 230 V a.c. V_3 230 V a.c. V_4

a   b   c   d

i For each transformer, state whether it is a step up or step down transformer.

....

....

[4 marks]

ii List the output voltages (V_1–V_4) in order from lowest to highest.

....

D–C

[1 mark]

5 Explain why you cannot use a transformer to change the voltage from a battery.

....

....

....

....

B–A*

[4 marks]

P5 Extended response question

This diagram shows a simple generator.
Describe how it produces electricity.

The quality of written communication will be assessed in your answer to this question.

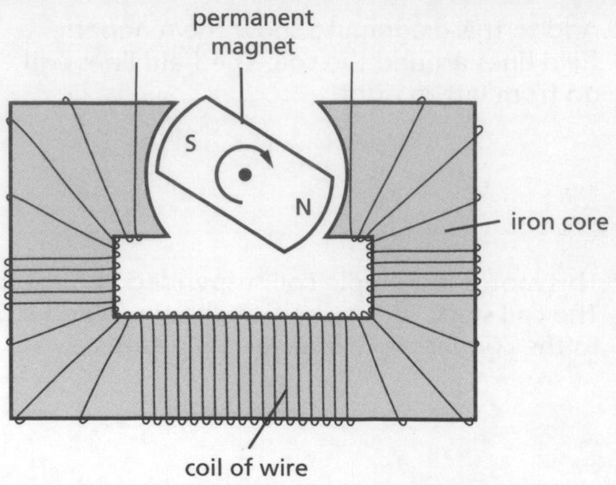

permanent magnet
S
N
iron core
coil of wire

(Source: OCR (A182) Twenty First Century Science - Physics A)

D–C

[6 marks]

Nuclear radiation

1 Complete the paragraph below about the structure of the atom with words from this list.

neutral   positively charged   mass   neutrons   protons

electrons   negatively charged   charge   nucleus

The atom has a .... containing neutrons and protons.

.... electrons orbit the nucleus. Overall the atom is neutral because

it has the same number of .... and ....

Most of the .... of the atom is concentrated in the nucleus, as the

electrons are very light.

[5 marks]

G–E

2 Rutherford and his team of scientists fired positively charged alpha particles at very thin gold foil, only a few atoms thick. The observations they made are shown below in the left-hand column. Draw arrows connecting the scientists' observations to the conclusions they made (in the right-hand column).

| **a** Most alpha particles went straight through the gold foil. |
| **b** Very few particles were deflected through small angles. |
| **c** Some particles bounced straight back. |

| **i** The nucleus is a very small but dense part of the atom. |
| **ii** The nucleus is positively charged. |
| **iii** Most of the atom is empty space. |

[3 marks]

D–C

3 Most carbon atoms are carbon-12. Carbon-14 is a radioactive isotope of carbon. What is an isotope?

....

....

....

[3 marks]

B–A*

4 Ionising radiation causes atoms to become ions. What is the difference between an ion and an atom?

....

....

....

....

[3 marks]

G–E

5 The pie chart opposite shows the various sources of background radiation.

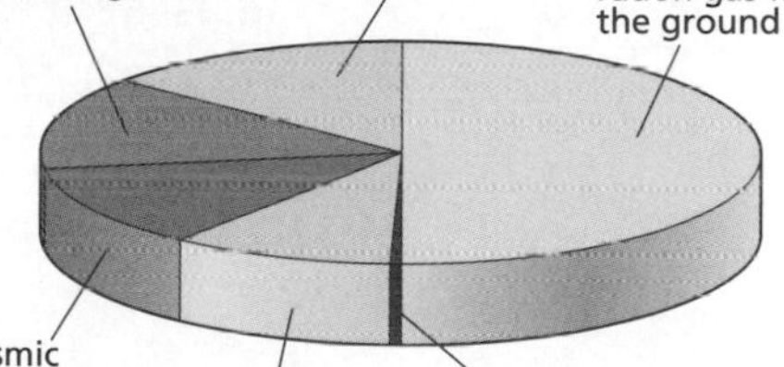

a What is meant by the term background radiation?

....

....

....   **[2 marks]**

gamma rays from ground and buildings 13.5%   medical sources 14%

radon gas from the ground 50%

cosmic rays 12%

food and drink 10%   nuclear power and nuclear weapons testing 0.5%

b Why do some parts of the country have more background radiation than others?

....

[1 mark]

D–C

6 Joanna and Michael are investigating the ionising radiation emitted from some rocks. Joanna suggests that they lower the temperature of the rocks because that will reduce the amount of radiation emitted. Michael disagrees with her – he thinks it will make no difference.

Discuss who you think is correct.

....

....

....

[3 marks]

B–A*

Types of radiation and hazards

1 There are three types of ionising radiation: alpha, beta and gamma.
For each description below, indicate which type of radiation is being described.

a This type of radiation is an electromagnetic wave. ....

b This type of radiation can be stopped by a piece of paper. ....

c This type of radiation is a large, positively charged particle. ....

d This type of radiation is not deflected by electric or magnetic fields. ....

e This type of radiation is a fast-moving electron. .... **[5 marks]**

2 Sunita and Caroline are trying to find out which type of radiation is emitted from a gas mantle. They set up the source and the detector in a line as shown opposite, and they have some sheets of paper, aluminium foil and thin steel sheets.

gas mantle   radiation detector

Describe how Sunita and Caroline can find out which type of radiation is being emitted.

....

....

....

....

.... **[5 marks]**

3 What is an alpha particle?

....

....

.... **[3 marks]**

4 The bar chart below shows the typical annual radiation dose for a person in Britain from different sources.

a What is the total radiation dose a typical person in Britain would get from ground and buildings and medical scans in one year?

(Source: OCR(A182) Twenty First Century Science – Physics A)

.... **[2 marks]**

b In 2011 an earthquake damaged Fukoshima nuclear power station in Japan, causing a radiation leak. Explain how the bar graph would be different for a person in Japan in 2011.

....

.... **[2 marks]**

5 Explain how ionising radiation inside the body causes mutations.

....

....

.... **[3 marks]**

Radioactive decay and half-life

1 Why does the amount of radiation emitted from a radioactive sample decrease over time?

....

G–E

....

[3 marks]

2 When uranium emits an alpha particle it changes into Thorium, and when carbon-14 emits a beta particle it becomes nitrogen.

Explain why elements that emit gamma rays do not become a different element.

D–C

....

....

[3 marks]

3 Polonium-216 undergoes alpha decay to form a radioactive isotope of lead (Pb).

Complete the balanced equation to show this decay.

$\frac{216}{84}$ **Po** ⟶ ☐☐ + ☐ **Pb**

B–A*

[5 marks]

4 Different radioactive sources have different half-lives. The graph opposite shows the activity over time for three different radioactive sources (A, B and C).

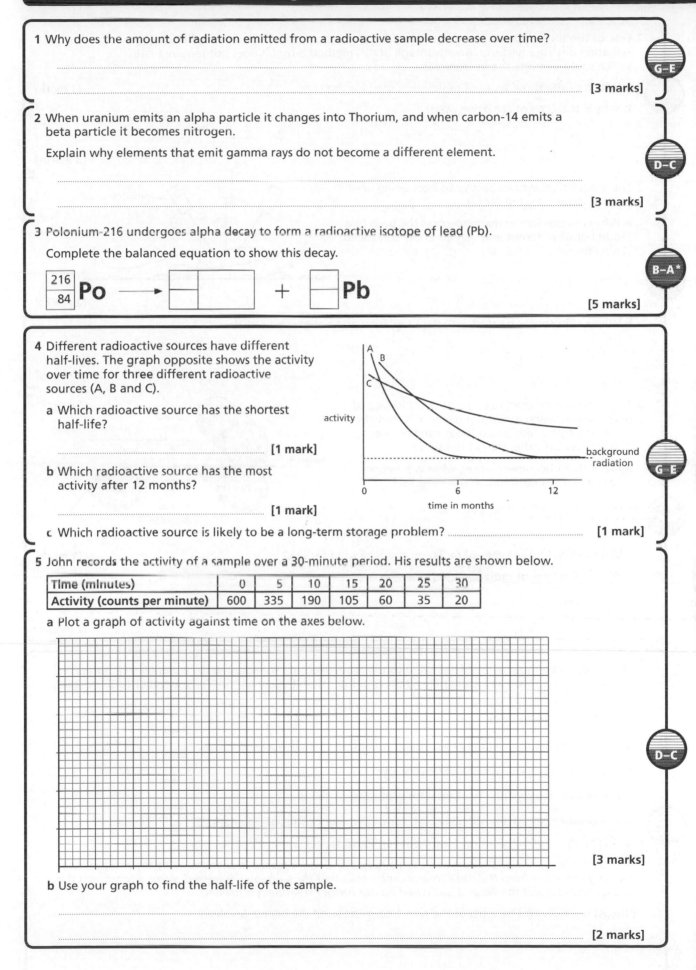

a Which radioactive source has the shortest half-life?

.... **[1 mark]**

b Which radioactive source has the most activity after 12 months?

.... **[1 mark]**

G–E

c Which radioactive source is likely to be a long-term storage problem? .... **[1 mark]**

5 John records the activity of a sample over a 30-minute period. His results are shown below.

Time (minutes)	0	5	10	15	20	25	30
Activity (counts per minute)	600	335	190	105	60	35	20

a Plot a graph of activity against time on the axes below.

[3 marks]

D–C

b Use your graph to find the half-life of the sample.

....

....

[2 marks]

Uses of ionising radiation and safety

1 One of the three types of radiation is used for sterilising medical instruments, because the radiation destroys bacteria. An advantage of this method is that it does not require heat, which could damage the instruments.

G–E

a Which of the three types of radiation is used for this? .... [1 mark]

b Why is this type of radiation used?

....

.... [3 marks]

2 The diagram shows radioactive sources being used to treat a deep-seated cancer.

radioactive source

a Why is the radiation made to enter the body in a number of different and very carefully controlled directions?

D–C

....

.... [2 marks]

b Why is skin cancer treated with beta radiation and not gamma radiation?

....

.... [2 marks]

3 Radiation can be used to monitor the thickness of paper as it is being made in a paper mill. Radiation is emitted by the emitter. It is detected by the detector on the other side of the sheet.

Source

Paper

Rollers

a If the sheet becomes thicker, what will happen to the level of radiation at the detector?

B–A*

Radiation detector

....

.... **[1 mark]**

b Which of the three types of radiation could be used for this? .... [1 mark]

c Why is this type of radiation used?

....

.... [2 marks]

4 The maximum annual risk of developing cancer from exposure to radiation for a worker in a nuclear reactor is 0.1%. This is approximately 40 times greater than the annual risk for a member of the public.

Why might this increased risk not be seen as a problem for the owners of the power station? Tick (✓) the correct answer.

a The owners are not required to consider the safety of their workers.

b The risk to a worker would still be very low.

c The owners supply their workers with protective clothing.

d The power stations are normally built far from major centres of population.

D–C

5 Read this description of a new treatment for breast cancer.

The cancer is cut out by the surgeon then a radioactive rod is placed in the wound by the radiographer. Ionising radiation from the rod kills any cancer cells that the surgeon has missed. After a few hours the rod is removed and the wound is stitched up. No further treatment is needed.

Discuss the risks and benefits of the new treatment to all the people involved.

....

....

....

.... [3 marks]

P6 Radioactive materials

Nuclear power

1 Complete the passage below using words from this list:

fusion   fission   bomb   fuel   power

About a sixth of the electricity produced in the UK is from nuclear power stations.

The nuclear .... is either uranium or plutonium. The energy

is released by nuclear ...., when a large nucleus splits into two

smaller nuclei. At the moment nuclear .... is not commercially

used to generate electricity.   **[3 marks]**

G–E

2 The difference in mass from fission reaction in a fuel pellet is 0.24 kg. Use Einstein's equation $E = mc^2$ to calculate the energy released from this reaction.

....

....   **[2 marks]**

B–A*

3 Give **two** advantages and two disadvantages of using nuclear fuels to produce electricity.

....

....

....

....   **[4 marks]**

D–C

4 In most nuclear power stations uranium is the nuclear fuel. For nuclear fission to occur a neutron must be fired at a uranium nucleus.

a Explain how this leads to a chain reaction.

....

....

....

....   **4 marks]**

b What is done in a power station to control the chain reaction?

....

....   **[2 marks]**

B–A*

5 In the Sun isotopes of hydrogen fuse together. What element is produced? Circle the correct response.

deuterium   helium   carbon   tritium   **[1 mark]**

G–E

6 Currently nuclear fusion is not commercially used to produce power.

a Explain **two** advantages that nuclear fusion has over nuclear fission as a source of power.

....

....   **[2 marks]**

D–C

b Describe the difficulties that must to be overcome before nuclear fusion can be used commercially.

....

....   **[3 marks]**

B–A*

A nuclear reactor produces radioactive materials for use in hospitals. The radioactive materials are used to treat patients.

Identify the different types of radioactive waste generated by the production and use of these radioactive materials, and describe how the waste should be dealt with.

The quality of written communication will be assessed in your answer to this question.

[6 marks]

The solar day

1 The Earth orbits the Sun. In which direction does the Sun appear to move across the sky?

.... [1 mark]

2 Astronomers talk about sidereal days and solar days.

a How long is a solar day?

.... [1 mark]

b Explain what astronomers mean when they talk about a solar day.

....

.... [1 mark]

3 Use the words provided to complete the sentences, you can use the word once or not at all.

Sun	Moon	stars	Earth	north
orbit	rotate	east	south	west

The Sun, Moon and .... all appear to .... around the

.... and travel .... to .... across the sky. [5 marks]

4 Explain the apparent motion of the stars and the Sun.

....

....

.... [2 marks]

5 Here are some statements about the apparent motion of astronomical objects and their position in the sky from one day to the next. Number them in order of shortest time to longest time.

a The time taken for the Moon to appear in the same position from one day to the next.

....

b The time taken for the Sun to appear in the same position from one day to the next.

....

c The time taken for the stars to appear in the same position from one day to the next.

.... [3 marks]

6 Write down the length of a sidereal day.

.... [1 mark]

7 By referring to the rotation of the Earth and the orbit of the Earth around the Sun explain why a solar day is longer than a sidereal day.

....

....

....

.... [3 marks]

The Moon

1 The following are statements about the phases of the Moon. Mark each statement as true or false by entering a T or an F in each box.

G–E

a We can see the Moon because it reflects light. ☐

b As the Moon orbits the Earth the amount of the lit side of the Moon we can see changes. ☐

c When we can see the entire lit side of the Moon this is called a new Moon. ☐

d The Moon orbits the Earth in just over 24 hours. ☐ **[4 marks]**

2 During which phase of the Moon does a solar eclipse occur? Tick (✓) the correct answer.

New Moon ☐

D–C

Half Moon ☐

Crescent Moon ☐

Full Moon ☐ **[1 mark]**

3 The diagram shows a solar eclipse. Match the descriptions **A, B, C, D** with the labels 1, 2, 3, 4 on the diagram.

A Umbra ....

B Penumbra ....

C Earth ....

D Moon .... **[3 marks]**

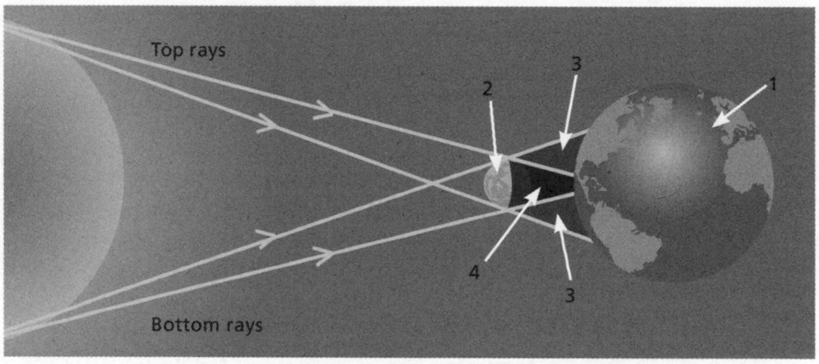

4 List **three** ways that a lunar eclipse is different from a solar eclipse.

....

....

....

.... **[3 marks]**

5 Eclipses only occur two or three times a year. By reference to the orbits of the Earth and the Moon, explain why they do not occur every full Moon and every new Moon.

B–A*

....

....

....

.... **[2 marks]**

The problem of the planets

1 List **three** factors that affect how easy it is to see the planets with the naked eye.

....

.... **[3 marks]**

2 Like the stars, the planets also appear to move east-west across the sky. What causes this apparent motion?

.... **[1 mark]**

3 Describe how the apparent motion of the planets differs from that of the stars.

....

....

.... **[2 marks]**

4 What words are used to describe the motion when planets appear to move backwards?

.... **[1 mark]**

5 Explain why a planet sometimes appears brighter and at other times appears dimmer.

....

....

.... **[2 marks]**

6 Mark the following statements as true or false.

All of the planets in our solar system orbit the Sun. ☐

The further away a planet is from the Sun, the longer its orbit takes. ☐

During its orbit the speed of a planet changes significantly. ☐

Sometimes planets change direction. ☐ **[4 marks]**

7 The diagram shows the apparent movement of Mars in the sky across a period of 35 weeks.

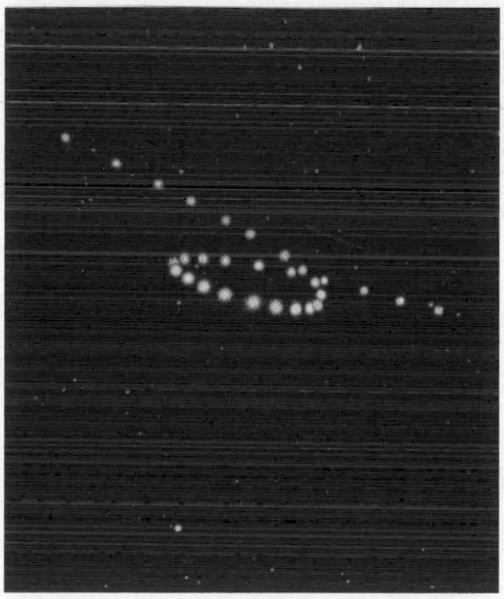

By referring to the orbit of the Earth and Mars about the Sun explain this retrograde motion.

....

....

....

.... **[4 marks]**

1 Use the diagram to help you explain why the stars that are visible in the night sky vary depending on the time of the year.

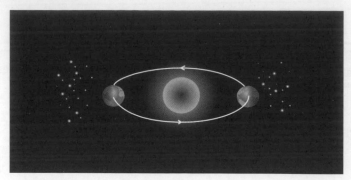

....

....

.... [2 marks]

2 During the summer in July the centre of the Milky Way will be overhead in the night sky. How would you expect the number of visible stars to be different than in January, when the night sky points in the opposite direction?

....

.... [1 mark]

3 The table below gives some information about stars.

Star	Declination	Ascension
Proxima Centauri	−62° 40′	14 h 29 m
Barnard's Star	+4° 41′	17 h 57 m
Altair	+8° 52′	19 h 52 m
Polaris	+89° 15′	2 h 31 m

a Which of the stars would be seen below the equator? .... [1 mark]

b Which of the stars is also called the North Star? .... [1 mark]

c Which **two** stars are you most likely to be able to see at the same time when looking

at the night sky? .... [2 marks]

4 What is meant by the ecliptic?

.... [1 mark]

5 Why is the ecliptic important to astronomers when using angles of ascension?

....

.... [1 mark]

6 What is the angle of declination?

....

.... [1 mark]

7 Why is the angle of ascension measured in hours, minutes and seconds instead of degrees, minutes and seconds?

....

.... [1 mark]

Refraction of light

1 What happens to a wave as it passes from one medium to another? Circle the correct answer.

its frequency changes   nothing happens   its speed changes   [1 mark]

G–E

2 How does the speed of light in a vacuum differ from the speed of light in glass?

.... [1 mark]

3 When waves hit a boundary between two materials at an angle the wave changes direction. What do we call this change in direction?

.... [1 mark]

4 Water waves travel more slowly in shallow water. On the diagram sketch the water waves as they cross the boundary.

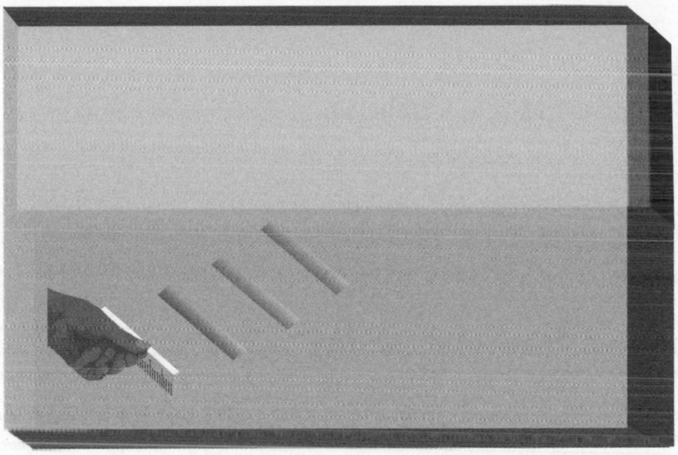

D–C

[2 marks]

5 On the diagrams below complete the paths of the rays of light as they pass through the glass blocks and out the other side.

[3 marks]

6 Lenses can make use of refraction to bring light to a focus. Describe the shape of a converging lens.

.... [1 mark]

7 The diagram below shows parallel rays striking a lens. Complete the path of the light rays as they pass through the lens and are brought to a focus.

B–A*

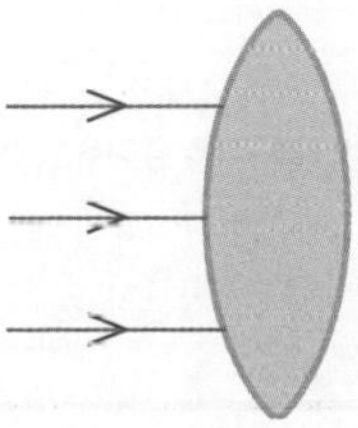

[4 marks]

Forming an image

1 Give a definition for the focal length of a lens.

....

.... [1 mark]

2 From the list below circle the factors that affect the focal length of a lens.

the temperature   **the material of the lens**

the curvature of the lens   **the diameter of the lens**   [2 marks]

3 The power of a lens is related to the focal length with the formula:

Power (dioptres) = 1/focal length (m)

G–E

a Lens A has a focal length of 5 cm. Calculate its power. Show your working.

....

.... [2 marks]

b Lens B has a power of 2 dioptres. Calculate its focal length. Show your working.

....

.... [2 marks]

c Both lenses in **a** and **b** are made of the same type of glass. Which of the lenses will
  have the greater curvature?

.... [1 mark]

4 Complete the passage below by selecting words from the list. You may use the words once,
  twice or not at all.

parallel   **principal axis**   **focal point**   **focus**   **distant**   **near**

D–C

Light rays from .... objects are effectively .... When a

lens focuses light from these objects the image will be formed at the ....

If the object is not in line with the centre of the lens the image will be formed off

the ....   [4 marks]

5 Complete the ray diagram below to show the formation of an image by a convex lens.

B–A*

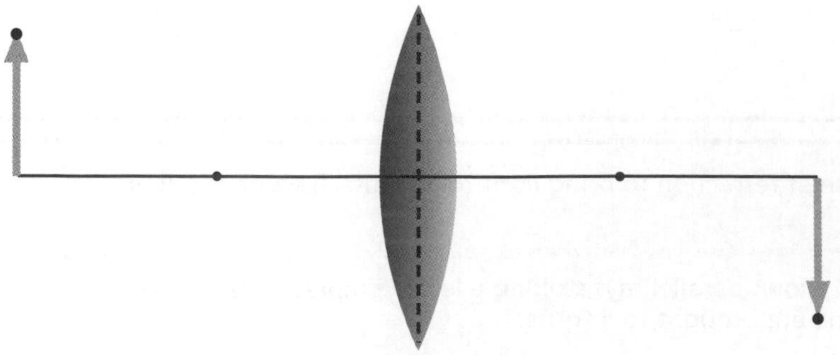

[4 marks]

6 Stars are seen as point objects. What kind of objects are galaxies, the Sun and planets
  in our solar system?

.... [1 mark]

The telescope

1 Mark the following statements with a T or an F to indicate whether they are true or false.

A simple refracting telescope contains two concave lenses. ☐

The objective lens has a long focal length. ☐

The eyepiece lens has a short focal length. ☐

The objective lens has a smaller diameter than the eyepiece. ☐

G–E

[4 marks]

2 A telescope is described as having an angular magnification of 3X.
What does this mean?

....

....

D–C

[2 marks]

3 The magnification of a telescope can be found by dividing the focal length of the objective lens by the focal length of the eyepiece lens.

a What is the magnification of a telescope that has a 2 m objective lens and a 10 cm eyepiece lens? Show your working.

....

....

....

....

[2 marks]

b What is the magnification of a telescope that has a 90 cm objective lens and a 5 cm eyepiece lens? Show your working.

....

....

....

....

B A*

[2 marks]

4 The magnification can also be found from the power of the lenses, using the formula:

$$\text{magnification} = \frac{\text{eyepiece power}}{\text{objective power}}$$

Carry out calculations to complete the table below.

	Magnification	Eyepiece power	Objective power
a	100		0.5 D
b		30 D	1.5 D
c	50	20 D	

[3 marks]

The reflecting telescope

1 Some telescopes use concave mirrors instead of objective lenses. State **two** advantages of using mirrors instead of lenses.

....

.... **[2 marks]**

2 The diagram shows a reflecting telescope. On the diagram label the eyepiece lens and the objective mirror. **[2 marks]**

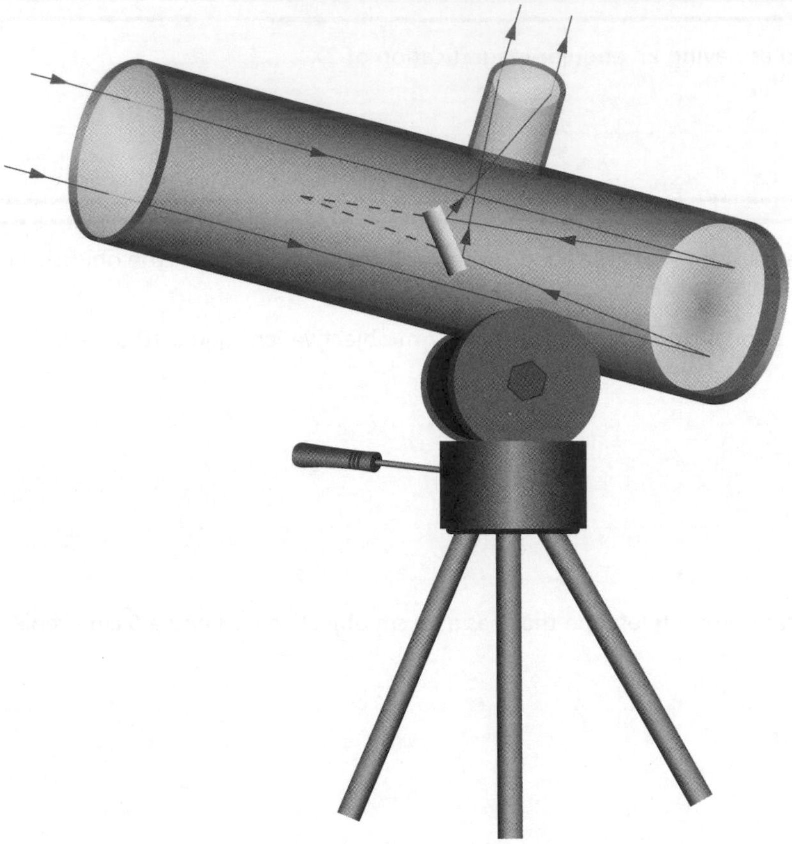

G–E

3 These sentences describe how an image is formed in a reflecting telescope; they are not in the correct order. Sort the sentences into the correct order.

a The objective mirror reflects the rays inwards towards a focal point.

D–C

b The eyepiece lens magnifies the image formed by the objective mirror.

c Parallel rays of light enter the telescope.

d A secondary mirror reflects the rays towards the eyepiece.

The correct order is: .... **[4 marks]**

4 Explain how lenses produce chromatic aberration.

....

....

.... **[3 marks]**

B–A*

5 Why is chromatic aberration a problem for astronomers?

....

.... **[1 mark]**

6 Concave mirrors do not produce chromatic aberration. Why not?

....

.... **[1 mark]**

P7 Further physics – studying the Universe

Diffraction

1 Give **two** reasons why telescopes are designed to have the objective lens as large as possible.

....

.... **[2 marks]**

2 As water waves pass through a gap they spread out.

What would you notice about how much they spread out if the gap was changed from a wide gap to a narrow gap?

....

.... **[1 mark]**

3 Why don't we notice light spreading out in the same way as it passes through a doorway?

....

.... **[1 mark]**

4 Countryside roads will often go into dips and behind hills but cars on them will still be able to detect long wavelength radio signals, even when they don't have a line of sight to the transmitter. This is not true for shorter wavelength mobile phone signals, which can be lost when the car goes into a dip between hills. Use diffraction to explain why.

....

....

....

.... **[4 marks]**

5 The three pictures show simulated photographs taken with the same telescope but with different aperture sizes.

A

B

C

a Sort the pictures into order of aperture size used, starting with the largest aperture.

.... **[1 mark]**

b Explain your answer to part a.

....

....

.... **[3 marks]**

6 The largest aperture telescopes in the world are radio telescopes. Use ideas of diffraction to explain why this is so.

....

.... **[2 marks]**

Spectra

1 State the **two** key bits of information astronomers can obtain by studying the spectrum of radiation emitted by a star.

....

.... [2 marks]

2 White light can be separated into the colours of the spectrum through the use of a prism. Use your understanding of refraction to describe how this takes place.

....

....

.... [2 marks]

3 Draw lines to connect the statements together to make **three** complete sentences about light shining through a prism.

| Red light | | is a mixture of wavelengths | | is refracted the least by the prism |

| White light | | has a short wavelength | | is refracted the most by the prism |

| Blue light | | has a long wavelength | | is split up by the prism |

[3 marks]

4 A diffraction grating is made of thousands of finely spaced gaps, which are normally around 500-600 nm wide. Why do the gaps in a diffraction grating have to be so narrow?

....

.... [2 marks]

5 A CD or a DVD can act like a reflection diffraction grating but give a different width spectrum.

Suggest why this is so.

....

.... [1 mark]

6 When the first quasar was discovered it was found that its spectrum was red-shifted so much that it must be 2.5 billion light years away. However it took 20 years for this to be accepted. Suggest why.

....

....

....

.... [2 marks]

The distance to the stars

1 Why is it difficult to tell how far away a star is by simply looking at how bright it appears?

....

.... **[1 mark]**

2 When astronomers use parallax to determine the distance to stars they look at the motion of nearby stars. What do they compare this motion to?

.... **[1 mark]**

3 The parallax method requires two measurements to be made at different times. If the first measurement was made in January, when would the second measurement be made?

.... **[1 mark]**

4 On the diagram, draw and label the parallax angle.

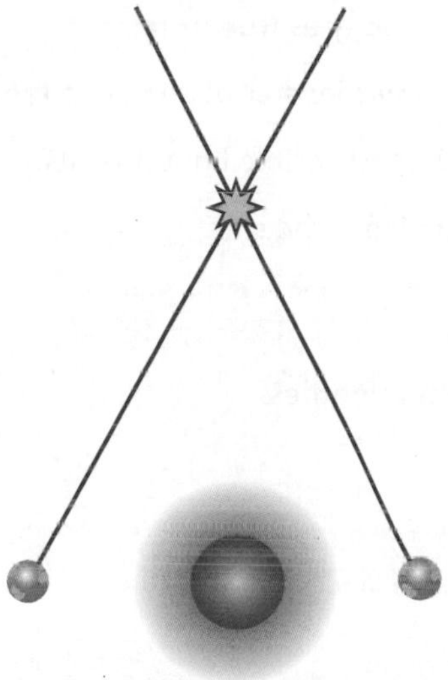

[2 marks]

5 Complete the table below.

Star	Distance (pc)	Parallax angle (arc seconds)
Sirius	2.64	
Arcturus		0.044
Barnard's Star		0.55
Proxima Centauri	1.30	

[4 marks]

6 A distant star has an observed parallax angle of 0.5 arcseconds. How far away is the star? Show your working.

....

.... **[2 marks]**

7 Explain why it is only possible to measure the distance to relatively close stars and not to distant galaxies.

....

.... **[1 mark]**

1 Give a definition of a light-year.

....

.... [1 mark]

2 Choose words from the list to complete the sentences below. You may use each word once or not at all.

G–E

| parsecs | light-years | kilometres | megaparsecs |

A parsec is around 3 .... The distance between one star and the next is normally

a few .... Intergalactic distances are measured in .... [3 marks]

3 Mark the following statements about luminosity as true or false.

D–C

A larger star will be more luminous than a smaller star of the same temperature. ☐

The luminosity of a star refers to just how much visible light it emits. ☐

The higher the temperature, the more luminous the star. ☐

Blue stars are cooler than red stars. ☐ [4 marks]

4 How does luminosity differ from apparent brightness?

....

.... [1 mark]

5 Explain what is meant by the inverse square law.

....

.... [2 marks]

6 A man stands 2 m from a lightbulb and measures its brightness to be 100 lumens. What measurements will he get if he stands:

a 1 m away

....

B–A*

b 4 m away

....

c 20 m away

.... [3 marks]

7 An astronomer studies two stars that have the same apparent brightness. If star A is a large blue star and star B is a small red star, state and explain which is further from Earth.

....

....

....

.... [3 marks]

Cepheid variables

1 What is meant by a Cepheid variable star?

....

.... **[1 mark]**

2 How is the period of a Cepheid variable star related to its luminosity?

....

.... **[1 mark]**

3 The graph shows the luminosity against period for Cepheid variable stars. The luminosity is measured in solar units; a value of 100 means the star is 100 times more luminous than the Sun.

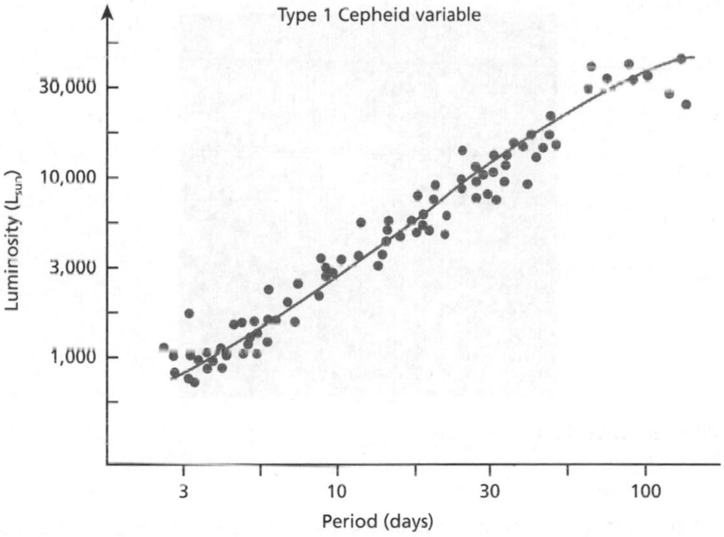

Use the graph to mark the statements below as true or false.

Cepheid variable stars are much more luminous than the Sun. ☐

A Cepheid variable with a period of 30 days is around 3000 times brighter than the Sun. ☐

A Cepheid variable with a period of 20 days will be twice as luminous as one with a period of 10 days. ☐ **[3 marks]**

4 An astronomer studies two Cepheid variable stars that appear to be of equal brightness. One has a period of 30 days and one has a period of 10 days. State and explain which is closest to Earth.

....

....

.... **[3 marks]**

5 Explain how knowing the luminosity of Cepheid variable stars has enabled astronomers to work out the scale of the Universe.

....

....

.... **[2 marks]**

Galaxies

1 Choose words from the list to complete the passage below.

the Milky Way   galaxy   the Sun   the Universe   stars

Our .... is called .... is a star in our galaxy,

which contains millions of .... and is one of millions of galaxies that

make up ....   **[5 marks]**

2 The development of telescopes led to the discovery of fuzzy bright clouds of light.

What name did astronomers give to these clouds?

.... **[1 mark]**

3 When astronomers first discovered these bright clouds there were a number of possible explanations.

  a Outline what is meant by the Island Universe Hypothesis.

  ....

  .... **[1 mark]**

  b Outline why some astronomers thought they were areas of planetary formation.

  ....

  .... **[1 mark]**

4 In the Curtis-Shapely debate of 1920 two prominent astronomers argued their ideas.
Connect the statements together to summarise the key points.

Heber Curtis	Believed that the clouds of light were spiral galaxies, like our own, but outside of our galaxy.	Believed that the clouds of light were small clouds of dust and gas in our galaxy.

Harlow Shapely	Believed that our galaxy was 300 000 ly across.	Believed that our galaxy was 30 000 ly across.

[2 marks]

5 Explain why the astronomers at the Curtis-Shapely debate did not reach a decision as to who was correct.

....

....

.... **[2 marks]**

The expanding Universe

1 How far across is the Universe?

.... [1 mark]

2 Edwin Hubble used a Cepheid variable star in the Andromeda nebula to prove that it was actually a distant galaxy. How did he prove this?

....

.... [2 marks]

G–E

3 Hubble studied Cepheid variable stars in many distant galaxies and found that almost all galaxies were redshifted.

  a What does this redshift show about the movement of the galaxies?

.... [1 mark]

  b How did this redshift also show that the Universe is expanding?

....

.... [1 mark]

  c What theory does this evidence support?

.... [1 mark]

D–C

4 The Hubble constant is used to show the link between the speed of recession of a galaxy and its distance from the Earth. Suggest why this constant has been refined many times since 1925.

....

....

.... [2 marks]

5 Complete the table below, which shows how the speed of recession and the distance from the Earth are linked for a number of galaxies (Hubble constant = 70 (km/s)/Mpc).

Galaxy	Speed of recession	Distance
NGC 4414		20 Mpc
Whirlpool	490 km/s	

[2 marks]

6 Measurements of the Hubble constant have been based on a wide range of galaxies.

  a Why is it important that the value of the Hubble constant is not based on observations of a single galaxy?

....

.... [2 marks]

  b The Andromeda galaxy has a recessional speed of -0.001 km/s and is a distance of 0.8 Mpc. What would the Hubble constant be if based on just this galaxy? Show your working.

.... [2 marks]

B–A*

The radiation from stars

1 How does the temperature of an object affect the radiation it emits?

....

.... [2 marks]

2 The diagram shows the connection between luminosity and frequency for three stars.

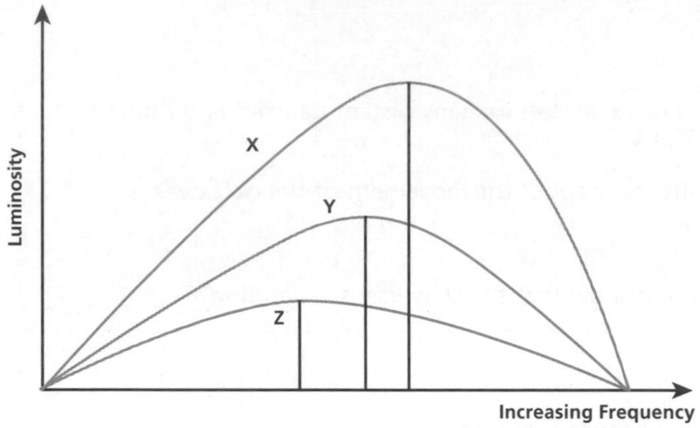

Mark each of the statements below as true or false by entering a T or an F in each box.

Star Z is probably a big star. ☐

Star Y is hotter than star Z. ☐

Star X has the highest peak frequency so it must be the hottest. ☐

Star Y is hotter than star X. ☐ [4 marks]

3 Antares is a star nearly 60,000 times more luminous than the Sun but it produces light of a lower frequency (longer wavelength). Explain how this is possible.

....

....

.... [2 marks]

4 Where in a star does nuclear fusion take place?

.... [1 mark]

5 What name is given to the part of the star that radiates energy into space?

.... [1 mark]

6 Use the following words to create a flow chart for the journey of energy that is produced in a star.

radiation zone   space   convection zone   core   photosphere

....

....

.... [3 marks]

Analysing stellar spectra

1 How does a continuous spectrum differ from a line spectrum?

....

.... **[2 marks]**

2 What would you see if you heated the gas of a single element and observed the spectrum it produced?

....

.... **[1 mark]** G–E

3 What would you see if you shone white light through this gas and observed the spectrum that was produced?

....

....

.... **[2 marks]**

4 What do dark lines in a star's spectrum tell astronomers about the star?

.... **[1 mark]**

5 The Sun contains mainly hydrogen and helium, but some large stars contain many other elements.

Explain what an astronomer will notice about the number of dark lines in the spectrum of a large star compared to the Sun. D–C

....

.... **[2 marks]**

6 Choose words from the list to complete the passage below. Words may be used once, twice or not at all.

| electrons | photons | atoms | elements |
| ions | energy | light | |

When .... lose .... they drop from one .... level to

another. When this occurs they emit .... of .... at a specific colour.

Because different .... have their own set of allowed levels different ....

have different emission spectra. **[7 marks]**

7 Use the ideas of absorption spectra to explain how a star's spectrum can tell us what elements it contains.

....

....

....

.... **[3 marks]**

Absolute zero

1 Here is a list of things that contain thermal energy.

   A. A burning match

   B. A bath of hot water

   C. A pan of boiling water

   D. An ice cube

a Sort the list into those with the highest thermal energy, most to least.

....

.... **[2 marks]**

b Sort the list into those at the highest temperature, most to least.

....

.... **[2 marks]**

2 If you heat up a gas, what happens to the speed of the molecules?

.... **[1 mark]**

3 What does the temperature of a gas depend on?

.... **[1 mark]**

G–E

4 What temperature is absolute zero in degrees Celsius?

.... **[1 mark]**

5 What unit do we measure absolute temperature in?

.... **[1 mark]**

D–C

6 This question is about gases and how the molecular model is used to explain their behaviour.

A gas is trapped in a piston but allowed to expand and contract so keeping the pressure constant. Explain, using a molecular model, what happens as the gas is heated.

....

....

....

.... **[3 marks]**

7 A sealed canister contains a gas at a pressure of 1 atmosphere. Use a molecular model to explain what happens to the pressure as the temperature is gradually reduced from 300 K until it reaches absolute zero. Include what the pressure will be at 150 K in your answer.

....

....

....

....

.... **[4 marks]**

B–A*

The gas laws

1 A student blows up a balloon and then ties it with a knot.

a Use a molecular model to explain how the pressure of the gas inside gives the balloon its shape.

....

....

.... [3 marks]

G–E

b How would the pressure inside the balloon change if the balloon is squeezed so its volume is reduced?

.... [1 mark]

2 State Boyle's Law.

....

.... [1 mark]

3 This question is about a fixed mass of gas at constant temperature. The gas starts at a pressure of 2 atmospheres and a volume of 6 m³. Use Boyle's Law to answer the following.

a What will the pressure be if the gas is compressed to a volume of 2 m³?

.... [1 mark]

D–C

b What will the volume of the gas be if the pressure is reduced to 1 atmosphere?

.... [1 mark]

4 A diver's air tank has a fixed volume of 5000 cm³ and is filled with compressed air at a pressure of 100 atmospheres at a temperature of 20 °C. Assuming that the air behaves as an ideal gas, answer the following questions.

a If the temperature increases to 40 °C, what will the pressure of the gas be?

....

....

.... [3 marks]

b After the tank is left in a cold place another diver measures the pressure and finds that it has reduced to 90 atmospheres. What temperature in degrees C will the compressed air be?

....

....

.... [3 marks]

B–A*

c When in use the compressed air is released at a pressure of 1 atmosphere. What volume of air is available for the diver to breathe?

....

....

.... [2 marks]

d The tank is rated to withstand a pressure of 200 atmospheres. At what temperature in degrees C would the pressure be so high that the tank would rupture?

....

....

.... [3 marks]

Star birth

1 What **two** elements are nebulae mainly comprised of?

....

.... **[2 marks]**

2 As gravity pulls these clouds of gas and dust together the temperature increases. Why does the temperature increase?

....

....

....

.... **[2 marks]**

3 Choose words from the list to complete the passage below. Words may be used once, or not at all.

    density   hotter   cooler   volume   pressure   temperature

In some areas of a nebula the .... is higher than others; this creates a stronger

gravitational attraction and means they can collapse inwards. As they collapse they become

.... and the .... increases. Eventually the ....

is so high that the cloud glows red and is known as a protostar. **[4 marks]**

4 A protostar is not an actual star.

What process needs to begin for a protostar to become an actual star?

.... **[1 mark]**

5 Mark the following statements as true or false by entering a T or an F in each box.

The Sun's energy comes from the combustion of hydrogen. ☐

The Sun is nearly 5 billion years old. ☐

Einstein theorised about how matter can be converted into energy. ☐

The Sun is powered by nuclear fission. ☐ **[4 marks]**

6 Explain why a protostar needs to increase in size before it becomes a star.

....

....

....

.... **[3 marks]**

P7 Further physics – studying the Universe

Nuclear fusion

1 What part of a star is the hottest?

.... [1 mark]

2 Despite the incredibly high density of a star, it is still a gas. Why is this so?

.... [1 mark]

3 What is released when two protons fuse together?

.... [1 mark]

4 Give a definition for an isotope.

....

.... [1 mark]

5 The following is the symbol for tritium (an isotope of hydrogen): $^3_1$H.

  a How many protons does tritium contain? .... [1 mark]

  b What is the mass number of tritium? .... [1 mark]

  c How many neutrons does tritium contain? .... [1 mark]

6 Look at this equation $^1_1$H + $^1_1$H ⟶ $^2_1$H + $^0_{+1}$e + $^0_0$v. The equation shows the fusion of two protons. Explain why the positron ($^0_{+1}$e) must be released in this reaction.

....

....

.... [2 marks]

7 Below are some equations for fusion reactions.

  x^4_2He ⟶ $^{12}_6$C

  $^1_1$H + $^{12}_6$C ⟶ $^Y_7$N

  What are the values of x, y?

  x = ...

  y = ... [2 marks]

8 State Einstein's energy equation.

....

.... [1 mark]

9 In the Sun four hydrogen atoms fuse to form one helium atom.

  By finding the mass lost, use Einstein's equation to calculate the energy released in Joules.

  Mass of hydrogen atom = 1.674×10^{-27} kg

  Mass of helium atom = 6.645×10^{-27} kg

  Speed of light = 3×10^8 m/s

....

....

....

....

.... [4 marks]

G–E

D–C

B–A*

The lives of stars

G–E

1 What **two** effects are balanced in a main sequence star?

....

.... [2 marks]

2 When will a star exit the main sequence phase of its life?

.... [1 mark]

D–C

3 Why is our Sun unable to form heavy elements but some stars can?

....

.... [2 marks]

4 What is the heaviest element that a star produces through energy-releasing fusion? .... [1 mark]

B–A*

5 The diagram shows a Hertzsprung-Russell diagram.

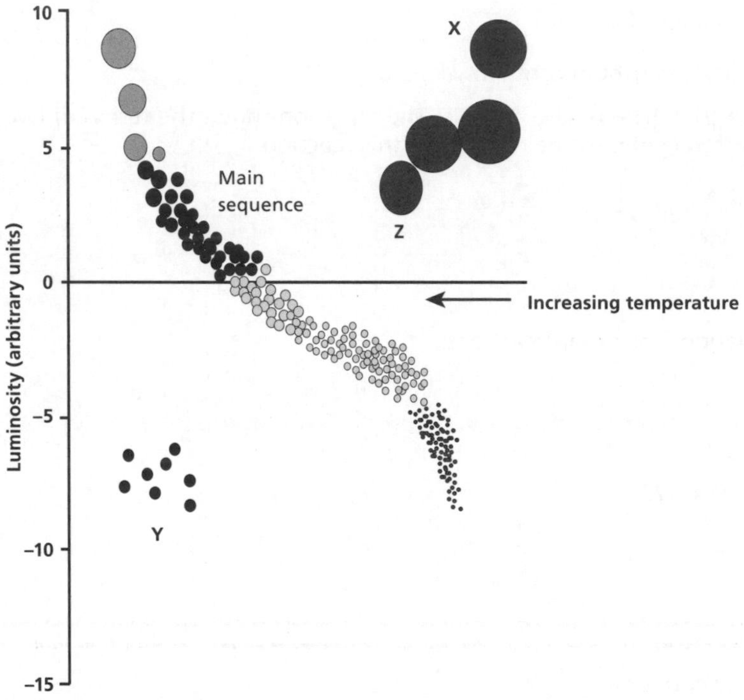

Which of the following statements are true? Put a tick (✓) in each box next to the **three** correct statements.

- **A.** Y is a white dwarf, Z is a supergiant and X is a red giant. ☐
- **B.** Y is a cold star because it is dim. ☐
- **C.** X is a supergiant, Y is a white dwarf and Z is a red giant. ☐
- **D.** Z is a bright and hot star. ☐
- **E.** The Sun would be approximately placed near the middle of the main sequence. ☐
- **F.** The hotter a main sequence star, the more luminous it is. ☐
- **G.** A star in the middle of the main sequence is hotter than a white dwarf. ☐ [3 marks]

6 How are white dwarf stars different to all other stars on the Hertzsprung-Russell diagram?

.... [1 mark]

The death of a star

1 Sort the following sentences into the correct order to explain the processes that take place at the end of a star's life.

As a star runs out of hydrogen...

A. Gravity is now able to crush the star.

B. Less energy is produced and the pressure drops.

C. The star collapses and the density and pressure increase once more.

D. The outer layer cools and the star becomes a red giant.

E. Fusion begins again and the star expands.

The correct order is: .... **[4 marks]**

D–C

2 This question is about nuclear fusion. The table gives the atomic masses of some elements.

Element	Hydrogen	Helium	Carbon	Nitrogen	Oxygen	Iron	Copper	Zinc
Atomic Mass (g)	1	4	12	14	16	56	64	65

a Which **two** elements in the table are parts of the fusion process in main sequence stars such as the Sun?

.... **[1 mark]**

b Which **two** elements in the table, apart from hydrogen, are not products of nuclear fusion in the core of a star?

.... **[1 mark]**

c Why can heavier elements like oxygen and iron be formed in the core of a red super giant but not in the core of a main sequence star such as the Sun?

....

.... **[1 mark]**

B–A*

3 Connect the following statements together to show what happens towards the end of a star's life, depending on the size of the star.

Sun-sized star	Expands to become a red giant or red supergiant.	Explodes in a supernova then contracts to become a neutron star.
Larger star	Expands to become a red giant.	Explodes in a supernova then contracts to become a black hole.
Largest stars	Expands to become a red supergiant.	Fusion stops, becomes a white dwarf and eventually black dwarf.

[3 marks]

4 Why does a star collapse when its core is mostly iron?

....

.... **[2 marks]**

The possibility of extraterrestrial life

1 Give **three** reasons why detection of extra-solar planets is difficult.

....

....

.... **[3 marks]**

2 An astronomer takes measurements of the observed brightness of three different stars across a period of several months. The results are shown in the graph below.

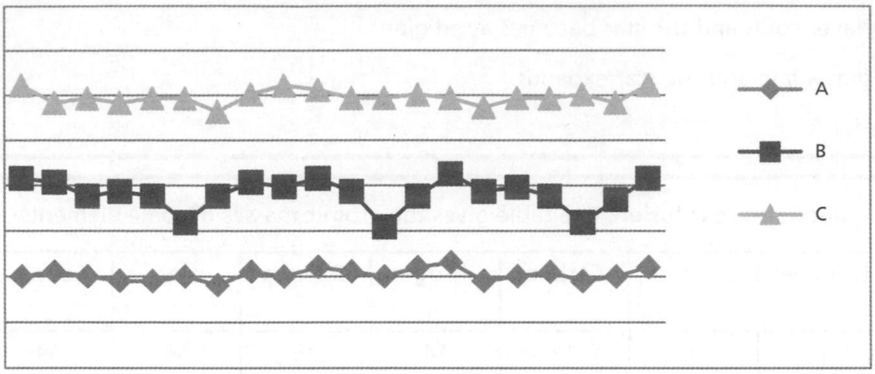

a Which star (A, B or C) is most likely to have an extra-solar planet orbiting it?

.... **[1 mark]**

b Explain your answer to part **a**.

....

....

.... **[2 marks]**

3 Explain what is meant by the Goldilocks zone and why it is important in the search for extraterrestrial life.

....

....

....

.... **[3 marks]**

4 State **one** place in our solar system where astronomers think they may find evidence of extraterrestrial life. Give a reason why they think they may find evidence there.

....

....

.... **[2 marks]**

Observing the Universe

1 Space-based telescope have a number of advantages and disadvantages.

From the list below tick (✓) the statements that correctly describe the advantages of space-based telescopes.

A. They are not affected by weather. ☐

B. They are expensive to set up, maintain and repair. ☐

C. They are not affected by refraction effects of the atmosphere. ☐

D. They can detect types of radiation that are absorbed by the Earth's atmosphere. ☐ **[3 marks]**

2 There are several locations of major astronomical observatories. At each site it is necessary to consider the factors that make it a suitable place to build a large telescope. One such site is the Mauna Kea Observatories in Hawaii.

a Suggest **two** astronomical factors that made this an ideal site.

....

.... **[2 marks]**

b Give **two** examples of economic or other factors that might also be important when planning and building an observatory.

....

.... **[2 marks]**

c A teacher explains that the telescopes in Chile are built well above sea level. A student then asks the teacher: 'Why can't an observatory be built in Scotland as it has mountains as well? Ben Nevis is over 1000 m above sea level.' Explain why this is not a good idea.

....

....

.... **[2 marks]**

3 The development of computer-controlled technology has provided many benefits to astronomers.

Describe **three** advantages of computer-controlled telescopes.

....

....

.... **[3 marks]**

International astronomy

1 Suggest **two** ways in which the siting of a large observatory might affect people living in the area.

....

....

....

....

....

.... **[2 marks]**

2 The Gemini Observatory in Chile was the result of shared work between Australia and six other countries.

  a Describe **two** benefits of sharing the work of setting up an observatory, rather than just one country being responsible.

....

....

.... **[2 marks]**

  b Suggest **one** thing that astronomers will need to agree after the telescope is complete.

....

.... **[1 mark]**

3 When astronomers work together on global projects they are able to make better progress than working by themselves.

Describe an example where astronomers from several countries, institutions or observatories have worked together towards a common goal.

....

....

....

....

.... **[3 marks]**

4 When countries work together to construct a telescope they will all expect to access it. State **three** ways in which astronomers could control the telescope.

....

....

.... **[3 marks]**

5 Outline **one** way that amateur astronomers are involved in cutting-edge astronomical research.

....

....

.... **[2 marks]**

P7 Extended response question

The Very Large Array (VLA) radio telescope consists of 27, 230-tonne, 25-metre diameter dish antennae that together comprise a single radio telescope system that has an affective aperture of over 30 km.

Explain the advantages and disadvantages of this type of telescope in terms of astronomical and non-astronomical considerations.

✏ *The quality of written communication will be assessed in your answer to this question.*

[6 marks]

P1 Grade booster checklist

I know that light travels through space at 300 000 km/s, and that a light-year is the distance travelled by light in a year.	
I know that we use light from distant stars and galaxies to detect them, and can measure distances to stars by comparing their brightness.	
I know that nuclear fusion is when two nuclei join together, forming a new element, and that this is the source of the Sun's energy.	
I understand that the Universe began about 14 thousand million years ago, and that distant galaxies are moving away from us.	
I can describe some rock processes taking place today and what they suggest about the past.	
I know that the Earth is about 4000 million years old.	
I know that continental land masses are moving very slowly and that this was described by Wegener in his theory of continental drift.	
I know that earthquakes produce P-waves and S-waves, and I can draw and label a diagram of the Earth's interior.	
I know that waves are disturbances that transfer energy. I can use the terms wavelength, frequency and amplitude, and can draw and interpret diagrams showing these.	
I am working at grades G/F/E	

I know the names, relative sizes and motion of different bodies in the solar system and the wider Universe, that the Sun is just one of the thousands of millions of stars in the Milky Way galaxy, and that the Universe contains thousands of millions of galaxies.	
I know how we can learn about distant stars and galaxies using their radiation, although light pollution and atmospheric conditions interfere with these observations.	
I know how relative brightness of stars and stellar parallax help us measure distances to stars.	
I understand that problems in measuring distances to and motion of distant objects mean that the future of the Universe cannot be accurately predicted.	
I can explain how Wegener's theory of continental drift was developed and modified; I understand that heating of the core causes convection in the mantle, and seafloor spreading.	
I know that earthquakes, volcanoes and mountain building generally occur at the edges of tectonic plates.	
I can describe the differences between P-waves and S-waves, and how they give evidence for the Earth's structure.	
I know how waves transfer energy, and can describe the difference between a transverse and longitudinal wave.	
I am working at grades D/C	

I know why the finite speed of light means we see distant objects in the Universe as they were in the past, and I understand why our observations of distant objects may be unreliable.	
I know that redshift shows more distant galaxies move away faster so space is expanding.	
I know that radiation emitted by the Earth has a lower principal frequency than radiation from the Sun, and that this radiation is absorbed or reflected back by some gases in the atmosphere.	
I can explain some problems with measuring the distances to and motion of distant objects and the mass of the Universe, and how this causes uncertainty in predicting its ultimate fate.	
I can explain the implications of the Earth being older than its oldest rocks.	
I know the implications of Wegener's theory and why geologists at first rejected it; I know the causes of seafloor spreading, and what the magnetisation of seafloor rocks can tell us.	
I know how moving tectonic plates cause earthquakes, volcanoes and mountain building.	
I understand how differences in P-waves (longitudinal waves) and S-waves (transverse waves) give evidence about Earth's structure.	
I am working at grades B/A/A*	

P2 Grade booster checklist

I know that a source emits electromagnetic radiation that is reflected, transmitted or absorbed by materials, and that it affects a detector when it is absorbed.	
I can list electromagnetic radiations in order of frequency and recall their speed through space.	
I know that absorbed electromagnetic radiation can heat and damage living cells.	
I know that some people worry about health risks from low-intensity microwave radiation.	
I know that ultraviolet radiation, X-rays and gamma rays are ionising radiation.	
I know that radioactive materials emit ionising gamma radiation continuously, and that exposure to ionising radiation can damage living cells, leading to cancer or cell death.	
I know that lead and concrete absorb X-rays, and that X-rays can produce shadow pictures.	
I know that some radiation from the Sun passes through the Earth's atmosphere, warming the Earth's surface.	
I know that carbon dioxide is a greenhouse gas and understand the causes of the greenhouse effect.	
I can interpret diagrams representing the carbon cycle.	
I know that global warming causes climate change and can describe some of these effects.	
I know that information superimposed onto an electromagnetic carrier wave creates a signal that can be transmitted.	
I know the features of analogue and digital signals, some advantages of digital signals over analogue signals, and that higher-quality sound or images use more digital information.	
I am working at grades G/F/E	

I know that energy from electromagnetic radiation is transferred by photons, and that higher frequency photons transfer more energy.	
I know that energy transferred by electromagnetic radiation depends on the energy and number of photons, and that electromagnetic radiation is less intense further from the source.	
I can relate the heating effect of radiation to the intensity of the radiation and its duration, and I know that water molecules strongly absorb microwave energy.	
I can explain why evidence for the health risk from microwaves is disputed.	
I know that photons of ionising electromagnetic radiations have enough energy to remove electrons from atoms or molecules when absorbed by substances.	
I know that sunscreen, clothing and the ozone layer absorb harmful ultraviolet radiation.	
I can apply understanding of the behaviour of X-rays to explain how images are produced.	
I know that all objects emit electromagnetic radiation, with a principal frequency that increases with temperature.	
I can use the carbon cycle to explain why the amount of carbon dioxide in the atmosphere was constant and is now increasing.	
I can explain and compare different ways in which electromagnetic radiation can transmit information.	
I know the advantages of digital signals over analogue signals, and that digital information is carried as pulses of an electromagnetic carrier wave, which are decoded when received.	
I am working at grades D/C	

I know that the intensity of electromagnetic radiation is the energy arriving per m^2 per s, and spreads over an increasing surface area and is partially absorbed as it moves away from the source.	
I know that ionised molecules can take part in chemical reactions.	
I know there are chemical changes in the atmosphere when ozone absorbs ultraviolet radiation.	
I know that radiation emitted by the Earth has a lower principal frequency than radiation from the Sun, and that this radiation is absorbed or reflected back by some gases in the atmosphere.	
I know that greenhouse gases include methane and water vapour, and that computer climate models provide evidence that human activities are causing global warming.	
I know that increased convection and water vapour in the hotter atmosphere can cause more extreme weather.	
I know why digital signals are less prone to noise than analogue signals, and how pulses of an electromagnetic carrier wave are created, are used to carry digital information and decoded.	
I am working at grades B/A/A*	

P3 Grade booster checklist

I know that power stations that burn fossil fuels emit carbon dioxide, and that growing energy demand raises issues about the availability and environmental effects of energy sources.	
I know that power is measured in watts, and that a more powerful appliance transfers energy more quickly. I can interpret information about energy use.	
I know that electric current passing through a component transfers energy to it and/or the environment, and a more efficient appliance transfers more of the energy to a useful outcome.	
I know that a domestic electricity meter measures energy use in kilowatt-hours (kWh).	
I can interpret and construct simple Sankey diagrams showing energy transfer.	
I can suggest how to reduce energy usage in personal and national contexts.	
I know that mains electricity is produced by generators, and that a generator produces a voltage across a coil of wire by spinning a magnet near it.	
I know that in many power stations a primary energy source heats water, producing steam which drives a turbine coupled to an electrical generator, and that some renewable energy sources drive the turbine directly. I can label a block diagram of the basic components of hydroelectric, thermal and nuclear power stations.	
I know that nuclear power stations produce radioactive waste, which emits ionising radiation.	
I know that electricity is convenient because it is easily transmitted and has many uses, and that the mains supply voltage to our homes is 230 volts.	
I know the advantages and disadvantages of different energy sources, and the effectiveness of methods of reducing energy use at home and work, and can interpret information about these.	
I am working at grades G/F/E	

I know that power in watts is the energy transferred each second, and can calculate how quickly an electrical device transfers energy using power = voltage × current.	
I can calculate energy transferred in joules or kWh using energy transferred = power × time.	
I can calculate the efficiency of an electrical device or power station using the equation efficiency = (energy usefully transferred ÷ total energy supplied) × 100%.	
I know how to calculate the cost of energy supplied by electricity.	
I can interpret and construct Sankey diagrams for various contexts, including electricity generation and distribution, and use them to calculate efficiency of transfer.	
I know how the voltage produced and current supplied by a generator can be increased, and that a generator uses more primary fuel per second when it supplies a bigger current.	
I can explain the difference between contamination and irradiation.	
I know how the distribution of electricity through the National Grid at high voltages reduces energy losses.	
I can discuss qualitatively and quantitatively the effectiveness of methods of reducing energy demand in a national context, and can interpret and evaluate information about different energy sources, considering efficiency, economic costs and environmental impact.	
I understand how different factors affect the choice of energy source for a given situation.	
I am working at grades D/C	

I know that power is the rate of energy transfer, and I can use and rearrange the equations power = voltage × current and energy transferred = power × time.	
I can use and rearrange the equation efficiency = (energy usefully transferred ÷ total energy supplied) × 100%.	
I can explain why contamination by a radioactive material is more dangerous than a short period of irradiation.	
I know that to ensure the security of national electricity supply, we need a mix of energy sources.	
I can interpret and evaluate information about different energy sources, also considering power output and lifetime.	
I am working at grades B/A/A*	

P4 Grade booster checklist

I know how to calculate the average speed of a moving object.	
I can draw and interpret distance–time graphs showing objects that are stationary or moving at constant speed.	
I know how to calculate the acceleration of moving objects.	
I know that forces between objects occur in 'interaction pairs', and I am able to identify the partner to forces.	
I understand that friction is a force between two moving surfaces, which acts to slow down the movement.	
I know that many forces can act on an object and I can find the resultant force.	
I know that falling objects accelerate towards Earth, but are slowed down by drag forces.	
I know about the safety features used in vehicles to reduce the forces in collisions.	
I understand that work is done when a force moves an object.	
I know the law of conservation of energy.	
I am working at grades G/F/E	

I understand the difference between the instantaneous speed and the average speed of a moving object.	
I know how to calculate the speed of an object from a distance–time graph.	
I can draw and interpret both distance–time graphs and speed–time graphs showing objects that are stationary, moving at constant speed and increasing speed.	
I know that when the forces on an object are balanced there is no acceleration of the object.	
I can give examples of situations where friction is wanted and unwanted.	
I can explain what is meant by terminal velocity.	
I can calculate the momentum of a moving object.	
I know that a resultant force causes a change in momentum.	
I know that energy is transferred when work is done.	
I can calculate the gravitational potential energy of an object.	
I know that in all energy transfers, some energy is always dissipated as heat energy.	
I am working at grades D/C	

I know what is meant by the terms displacement and velocity.	
I understand that acceleration can cause a change in direction as well as a change in speed.	
I can calculate acceleration from speed–time graphs.	
I can explain friction and reaction forces.	
I can draw force diagrams.	
I can explain how the forces on moving vehicles affect their speed.	
I can calculate the forces involved when there is a change of momentum.	
I can calculate the kinetic energy of moving objects.	
I am working at grades B/A/A*	

P5 Grade booster checklist

I know that rubbing insulators causes them to become charged, and that like charges repel and unlike charges attract.	
I know that electrons are negatively charged.	
I know that electric current is a flow of charge and is measured in amps.	
I know that the larger the voltage in a circuit the higher the current, and the larger the resistance in a circuit the lower the current.	
I know that LDRs are semi-conductors whose resistance changes with light intensity, and that thermistors are semi-conductors whose resistance changes with temperature.	
I understand current and resistance in series circuits.	
I know that electromagnetic induction produces a voltage, and that a current will flow if there is a circuit.	
I can describe a transformer.	
I know the difference between d.c. and a.c., and I know that the domestic electricity supply in the UK is 230 V a.c.	
I know that electric motors are used in many different appliances.	
I am working at grades G/F/E	

I know that metals (conductors) contain lots of electrons which are free to move, and that when a battery is connected in a circuit it causes the free electrons to move around the circuit.	
I understand that work is done by a battery when it causes an electric current to flow, and that energy is transferred to the component (usually as heat).	
I can calculate electrical power.	
I can calculate resistance and I know how the current in a resistor varies as the voltage is increased.	
I know that the potential difference (or voltage) is a measure of the work done by the charges in a circuit.	
I can explain currents in parallel circuits, and potential differences in series circuits.	
I understand that if a wire experiences a changing magnetic field, a voltage will be induced.	
I can describe how a generator (or dynamo) works and explain how you can increase the size of the voltage produced.	
I know that a wire carrying a current in a magnetic field will experience a force, and that this is the principle of the motor effect.	
I am working at grades D/C	

I can explain why resistance changes with temperature in metals.	
I can explain currents and potential differences in series and parallel circuits.	
I can explain clearly how a transformer works, and know that the voltage ratio equals the number of turns ratio.	
I can explain why a.c. is used for domestic electricity.	
I can explain how motors work.	
I am working at grades B/A/A*	

P6 Grade booster checklist

I know that radioactive elements emit ionising radiation, and that some radioactive elements are naturally occurring and contribute to background radiation.	
I know that an atom has a positively charged nucleus, made of protons and neutrons, which is surrounded by electrons.	
I know that the behaviour of radioactive materials cannot be changed by chemical or physical processes.	
I can describe the penetration properties of the three types of ionising radiation (alpha, beta, gamma).	
I understand that over time, the activity of radioactive materials decreases.	
I know that ionising radiation can damage living cells and may cause cancer, and that it is used to treat cancer, sterilise food and surgical instruments, and as a tracer.	
I know that radiation dose (in Sieverts) is a measure of the possible harm done to your body, and that people are exposed to risk by contamination or irradiation.	
I know that nuclear fission and fusion can produce a lot more energy than chemical reactions.	
I understand that nuclear power stations use nuclear fuels, such as uranium, and produce radioactive waste.	
I know that some people, such a radiographers and nuclear power workers, are regularly exposed to radiation, and that their exposure to radiation must be monitored.	
I am working at grades G/F/E	

I understand that the results from the alpha-scattering experiment give evidence for the structure of the atom.	
I know that two hydrogen nuclei can fuse to form helium in a nuclear fusion reaction.	
I know the relative mass and charge of alpha and beta particles, and can relate this to their ionising power.	
I understand the half-life of radioactive materials, and know that there is a very wide range of values for half-life.	
I know that ionising radiation causes atoms to become ions, and I can list some uses of ionising radiation.	
I can interpret data on risk related to radiation dose, and know that we are exposed to radiation all the time from background radiation.	
I can relate half-life to the amount of time it takes for a radioactive source to become safe.	
I understand the difference between nuclear fission and fusion reactions.	
I know that radioactive waste is categorised as high level, intermediate level and low level, and I can relate this to disposal methods.	
I am working at grades D/C	

I know that protons and neutrons are held together in the nucleus by a strong nuclear force, which overcomes the electrostatic repulsion of the protons.	
I know that Einstein's equation $E = m c^2$ is used to calculate the energy released during nuclear fission and fusion reactions.	
I can explain what isotopes are.	
I know that an alpha particle is a helium nucleus (2 protons and 2 neutrons), and that a beta particle is a fast-moving electron.	
I can complete nuclear equations for alpha and beta decay, and I know that radioactive nuclei are unstable. I can carry out simple calculations involving half-life.	
I know that ions formed by ionising radiation can take part in chemical reactions.	
I can explain the uses of ionising radiation, using ideas about the properties of the ionising radiation and half-life.	
I understand that nuclear fission involves a neutron hitting a large unstable nucleus, which then splits into two smaller nuclei, releasing neutrons and energy.	
I understand that nuclear fission needs to be controlled in nuclear power stations, and can use the terms chain reaction, fuel rod, control rod and coolant.	
I am working at grades B/A/A*	

I know what a solar day is.	
I can explain the phases of the Moon.	
I can recall the names of the naked eye planets and explain why we can see them without a telescope.	
I can explain why different stars are visible at different times of the year.	
I can explain what is meant by refraction.	
I can draw ray diagrams showing how a convex lens focuses parallel rays of light.	
I can calculate the power of a lens.	
I can describe the construction of a simple telescope.	
I can recall the differences between a reflecting and a refracting telescope.	
I can explain what is meant by diffraction.	
I can describe how a prism can split white light into different colours.	
I can explain what is meant by parallax.	
I know about light-years, parsecs and mega parsecs and can explain when these would be used.	
I know what Cepheid variable stars are.	
I know what the Milky Way is.	
I can describe how Hubble measured the distance to galaxies.	
I can describe the radiation emitted by hot objects.	
I know what is meant by absorption and emission spectra.	
I can relate temperature to the kinetic energy of molecules.	
I know that at absolute zero the molecules of a gas do not have any kinetic energy.	
I can explain what happens to the pressure of a gas as the volume is changed.	
I know what a protostar is and where they are formed.	
I can describe how energy is produced in a star.	
I can describe the main sequence of a star.	
I can describe two methods for detecting extra-solar planets.	
I can explain the advantages and disadvantages of space-based telescopes.	
I can describe astronomical and non-astronomical factors that should be considered when deciding on siting new observatories.	
I am working at grades G/F/E	

P7 Grade booster checklist

I can explain the apparent motion of the Moon and stars.	
I can describe the formation of solar and lunar eclipses.	
I can describe the motion of the planets.	
I can describe how angles of declination and right ascension are used to find the position of a star.	
I can explain why waves change direction when passing from one medium to another.	
I can describe how a convex lens focuses light from the stars.	
I can explain what is meant by the magnification of a telescope.	
I can describe the advantages of the reflecting telescope.	
I can describe the factors that affect the amount a wave is diffracted.	
I can explain how a diffraction grating can split white light into different colours.	
I can explain the parallax method for calculating the distance to stars.	
I know what is meant by the luminosity of a star.	
I know how the period of a Cepheid variable star is related to its luminosity.	
I can describe what nebulae are.	
I can describe how redshift in the spectra of Cepheid variable stars led to the discovery that the galaxies were moving away from the Earth.	
I can describe how the surface temperature of a star is linked to its luminosity and the peak frequency of the emitted radiation.	
I can describe how the spectrum of a star can provide information about its composition.	
I understand what is meant by absolute zero and can convert from °C to Kelvins.	
I can use a molecular model to explain temperature and pressure.	
I can explain Boyle's Law.	
I can describe the formation of a protostar.	
I can use nuclear equations to describe nuclear fusion reactions.	
I can explain why small stars do not make heavy elements.	
I can describe what will happen to the Sun when its hydrogen runs out.	
I can describe what is meant by the Goldilocks zone.	
I can describe the problems of ground-based telescopes.	
I can explain, with examples, why modern astronomy needs international cooperation.	
I am working at grades D/C	

I can explain the difference between a sidereal day and a solar day.	
I can use the relative orbits of the Sun, Moon and Earth to explain why eclipses do not occur every month.	
I can explain the retrograde motion of the planets.	
I understand what is meant by the celestial sphere and how it is used to describe the position of astronomical objects.	
I can use refraction to explain how a lens brings rays of light to a focus.	
I can draw ray diagrams to explain how a convex lens forms an image of an extended object.	
I can calculate the magnification of a telescope.	
I can explain the cause of chromatic aberration and why a reflecting telescope prevents this.	
I can explain the effect of diffraction on telescope images and how this links to the aperture size.	
I can use the parallax angle to calculate the distance to a star.	
I can use the inverse square law to explain how the distance affects the apparent brightness of a star.	
I can explain the importance of Cepheid variable stars in mapping the scale of the Universe.	
I can outline the key features of the Curtis-Shapley debate.	
I can make calculations using Hubble's Law.	
I can explain how Hubble's results provided evidence for the expanding Universe.	
I can explain how energy is transported through a star.	
I can use the idea of energy levels to explain emission and absorption spectra.	
I can use the gas laws to carry out calculations of the pressure, volume and temperature changes of an ideal gas.	
I can use Einstein's energy equation to calculate the energy released during nuclear fusion.	
I can use the Hertzsprung-Russell diagram to identify stars of different types.	
I can explain what happens to different sized stars at the end of their lives.	
I understand that there is currently no evidence for life existing anywhere else in the Universe.	
I can discuss the advantages of computer-controlled telescopes.	
I am working at grades B/A/A*	

Notes

Notes

Answers

P1 The Earth in the Universe

Page 86 Our solar system and the stars

1 Comet, star, planet, moon, dwarf planet

2 a Mercury; Earth; Uranus; Jupiter

  b Mercury; Earth; Jupiter; Uranus

3 Light has a speed of 300 000 km/s; a light-year is how far light goes in a year; this is $300\,000 \times 60 \times 60 \times 24 \times 365 = 9.46 \times 10^{12}$ km

4 The Milky Way is one of many *galaxies* in the Universe. Each galaxy is made of many *stars*

5 Look at radiation from stars; and compare it with radiation from the Sun. If they are different, then hypothesis is disproved

6 Radiation from a distant galaxy takes a long time to reach us; but it can only tell us about the galaxy when it left. So if the galaxy is a billion light-years away, we see it as it was a billion years ago

7 Earth; Sun; Solar System; Milky Way

8 Stars which are further away have a lower apparent brightness than ones which are closer; so the relative brightness of two stars allows you to work out their relative distance; if you know the distance to one star, then you can work out the distance to the other

9 Only close stars have a large enough parallax to be measured accurately; most stars do not appear to move relative to the others as the Earth moves around its orbit

Page 87 The fate of the stars

1 The energy comes from fusion of hydrogen

2 Nuclear fusion in stars; forces hydrogen atoms together to make helium. Big enough stars can use fusion to make helium into heavier elements. The heaviest elements are made when a star explodes at the end of its lifetime

3 Solid planets are made out of elements heavier than hydrogen or helium. Scientists believe that when the Universe was created it only contained hydrogen and helium. Other elements were only created when the first stars ran out of fuel and exploded as supernovae. The dust and gas from these explosions then condensed to make our planets

4 Most galaxies are moving away from us. This increases the wavelength of the light we receive from them

5 The Sun is 5000 million years old; the Earth is 4500 million years old; the Universe is 14 000 million years old

6 The future expansion of the Universe depends on its mass. If there is enough mass in the Universe, gravity will eventually reverse the expansion and the Universe will end up at a single point. If there is too little mass, the Universe should continue to expand forever. But finding the mass of the Universe is difficult because we can only see those bits of the Universe which emit radiation

Page 88 Earth's changing surface

1 Fossils are the remains of plants and animals in rock. Mountains are made by folding the rocks of the Earth's crust. Volcanoes are new mountains made from lava. Sediments are materials from the erosion of mountains

2 Mountains are broken down into small bits by the weather. The bits are carried into the sea where they fall to the bottom and form layers of sediments. Each new layer buries and crushes the previous layers to make rock

3 The Earth's crust is always moving. Where bits of the crust collide head-on, the sedimentary rocks made under the sea from material eroded from mountains are folded and pushed up to make new mountains. Where bits of the crust are moving apart, liquid rock can rise up from the centre of the Earth to make new mountains called volcanoes out of lava. So although erosion is always scraping the tops off mountains and putting them into the sea, new mountains are made all the time to replace them

4 b; e

5 Wegener's theory was very different to the other theories which explained similar rocks and fossils where the continents fitted together. Wegener had no science qualifications. Nobody could detect the motion of the continents. Nobody could explain why the continents should move

6 Lava comes up from the mantle at the oceanic ridge down the centre of the seafloor. As the lava solidifies, it is magnetised by the Earth's field and then gradually moves away from the oceanic ridge as it makes new rock. Every so often the Earth's magnetic poles swap over, changing the magnetisation direction of the new rock made at the ridge. So magnetisation of the rocks suddenly changes as you move away from the ridge

Page 89 Tectonic plates and seismic waves

1 Earthquakes; volcanoes; mountains

2 Where two tectonic plates meet. Tectonic plates are pieces of the Earth's crust

3 Liquid magma comes up where plates are moving apart, forming volcanoes. Where plates are moving together, one plate may be forced under the other. The plate melts as it moves down; and the liquid rock is forced up through cracks in the other plate to form volcanoes

4 As the plates move past each other, tension built up at the boundary is suddenly released as an earthquake

5 Most of the original rocks were eroded and formed sediments in the sea. Some of these sediments were pushed into the mantle and melted; as one tectonic plate was pushed under another. The material was then fed back up to the surface through volcanoes to make fresh rock

6 Core – liquid iron; crust – solid rock; mantle – semi-liquid rock

7 Tectonic; seismic; S-waves; P-waves; faster

8 The evidence comes from observations of seismic waves; around the world. The time of arrival of the waves allows the speed of the waves to be calculated. This tells us about the density of the material that the waves have passed through. There are places where S-waves don't arrive, suggesting a liquid core which S waves can't pass through

Page 90 Waves and their properties

1 Speed – how far the energy of the wave travels in a second; Frequency – the number of vibrations of the wave source in one second; Amplitude – maximum value of the disturbance in one wave; Wavelength – distance along wave from one zero disturbance to the next

2 Energy

3 Time delay for light is almost nothing, so distance $= 300 \times 1.5 = 450$ m

4 The frequency of a wave is the number of vibrations per second produced by it

5 Speed – m/s; frequency – Hz; wavelength – m

6 Frequency $= 20\,000$ Hz, wavelength $= 0.40$ m; Speed $= 20\,000 \times 0.40 = 8\,000$ m/s

7 a Frequency $=$ speed / wavelength $= 2.0 \times 10^8 / 0.44 \times 10^{-6}$ $= 4.5 \times 10^{14}$ Hz

  b Assuming speed is still 2.0×10^8 m/s; wavelength $=$ speed / frequency $= 2.0 \times 10^8 / 6.7 \times 10^{14} = 3.0 \times 10^{-7}$ m

Page 91 P1 Extended response question

5–6 marks

Answer includes the majority of relevant points for both theory and observations, as follows:

- Relevant scientific points about theory of an expanding Universe:
  - Every galaxy should appear to be moving away from every other galaxy.
  - This should redshift the light from each galaxy.
  - Redshift increases the wavelength of the light from it received by another galaxy.
  - The further away the galaxies are, the greater the redshift should be.
- Relevant scientific points about observations of the Universe:
  - Observations of the spectra; of galaxies seen from Earth shows a redshift.
  - A few nearby galaxies do not have a redshift.
  - Fainter galaxies have a larger redshift than brighter ones.
  - This suggests that redshift increases with distance between galaxies.

All information in answer is relevant, clear, organised and presented in a structured and coherent format. Specialist terms are used appropriately. There are few, if any, errors in grammar, punctuation and spelling

3–4 marks

Answer includes at least half of the relevant points (see above) for both theory and observations. For the most part the information is relevant and presented in a structured and coherent format. Specialist terms are used for the most part appropriately. There are occasional errors in grammar, punctuation and spelling

1–2 marks

Answer includes at least half of the relevant points (see above) from either theory or observation. Answer may be simplistic. There may be limited use of specialist terms. Errors of grammar, punctuation and spelling prevent communication of the science

0 marks

Insufficient or irrelevant science. Answer not worthy of credit

P2 Radiation and life

Page 92 Waves which ionise

1 b; a; d; c

2 300 000 km/s

Answers

3 Radio waves ⟶ infrared ⟶ ultraviolet ⟶ gamma rays

4 Energy; photons; light; frequency

5 b; c; d

6 The intensity of a wave decreases as it moves away from its source. This is because the photons are spread over an increasing area as the wave moves

7 Intensity = energy / (time × area) = 4.8×10^{-3} / ($8.0 \times 2.0 \times 10^{-6}$) = 300 J/s/m^{-2}

8 The atom becomes positively charged. This is because the electron which was removed has negative charge, and the atom started off with no charge at all

9 a Gamma rays; X-rays; ultraviolet

  b The photons have enough energy; to knock an electron out of the atom

10 Photons absorbed by a molecule in a cell; can ionise it; starting off a chemical reaction; which might damage a cell's DNA; and cause cancer

Page 93 Radiation and life

1 a ii

  b iii; iv

2 Absorbed; transmitted; shields; decreasing

3 a; b; c

4 The microwaves are only absorbed; by water molecules in food

5 They select a large number of people who use mobile phones; and another control group of people who don't; but are otherwise the same. They then record how many people in each group develop cancer; over many years

6 a Ultraviolet

  b Give you skin cancer

  c Put on sunscreen; and cover up with clothes

7 a You can get sunburn; and skin cancer

  b You get a tan; and it improves your health (by making vitamin D)

8 As the ozone absorbs the radiation; it is chemically changed

Page 94 Climate and carbon control

1 b; f

2 All; greatest; increasing

3 A – ii; B – i; C – iii; D – iv

4 Plants have absorbed carbon through photosynthesis; at the same rate; as it has been released by respiration; of living organisms

5 Methane has the most effect; then carbon dioxide; and finally water vapour. There is a lot of water vapour, so it has most effect. There is very little methane, so its effect is small

6 Farmland could be flooded by the sea; it could become too hot for some crops; the weather could become too stormy

7 The model is a computer program; which uses laws of science; to work out future climate from today's climate. A good model will be able to predict today's climate from that of some period in the past

Page 95 Digital communication

1 Radio, microwaves, visible and infrared

2 They are not absorbed by air; so can travel from the transmitter to receiver

3 It can travel through optical fibres; with very little absorption by the glass

4 It changes either the frequency; or the amplitude; of the wave

5 C

6 a They are the two values for the carrier wave; which can be on or off

  b The value of the sound is coded as a string of 1s and 0s; many time a second; at the transmitter. The strings are used to switch the carrier wave on and off. The receiver uses the strings of code to construct a copy of the original sound

7 The noise signal is there all the time; so can be usually distinguished from the digital signal; which turns the carrier wave on and off

8 2; 1; 0; 8

9 The player receives a large number of samples in each second. They arrive one after the other. Each sample is a string of binary digits; which is used to set the value; of the sound at that instant

10 The information is easily stored electronically; it can be processed in a computer; it can be music, speech or picture; it doesn't get lost or altered during transfer

Page 96 P2 Extended response question

5–6 marks

Answer includes majority of the following points:

• Relevant scientific points about carbon cycle for trees:

  – Trees absorb atmospheric carbon through photosynthesis.

  – Carbon is stored as wood.

  – Once the tree has been cut down it is eaten by organisms.

  – Respiration of organisms returns carbon to the atmosphere.

• Relevant scientific points about results of removing trees:

  – Replacement vegetation may not absorb as much atmospheric carbon.

  – Less wood to act as long-term store of carbon.

  – Soils may become degraded and less fertile so there will be less photosynthesis.

  – Soil may be washed away resulting in less photosynthesis.

All information in answer is relevant, clear, organised and presented in a structured and coherent format. Specialist terms are used appropriately. There are few, if any, errors in grammar, punctuation and spelling

3–4 marks

Answer includes at least half of the points from both lists (see above). For the most part the information is relevant and presented in a structured and coherent format. Specialist terms are used for the most part appropriately. There are occasional errors in grammar, punctuation and spelling

1–2 marks

Answer includes at least half of the points from one of the lists (see above). Answer may be simplistic. There may be limited use of specialist terms. Errors of grammar, punctuation and spelling prevent communication of the science

0 marks

Insufficient or irrelevant science. Answer not worthy of credit

P3 Sustainable energy

Page 97 Energy sources and power

1 Coal, oil and gas

2 Biofuels; wind; solar; waves; geothermal; hydroelectric; tidal *(Any 3)*

3 It has to be transferred from another source of energy

4 Most of our electricity comes from burning fossil fuels; and these are non-renewable so they will run out one day

5 Burning gas makes carbon dioxide. This is a greenhouse gas; so it increases the temperature of the atmosphere. This impacts on the environment by causing floods and stormy weather

6 Energy; watts; joules; seconds

7 250 W is 0.25 kW, so energy = 0.25 × 24 = 6.0 kWh

8 The energy in the circuit – flows from the supply to the kettle

  The current in the circuit – transfers energy from the supply

  The power of the kettle – is the energy transfer per second

  The voltage of the supply – provides the current with energy

9 Power; volts; current; amperes

10 230 × 3.0 = 690 watts

11 Power = 1150 W so current = 1150 / 230 = 5.0 A

Page 98 Efficient electricity

1 One kilowatt-hour is – 3 600 000 joules. The meter readings are in – units of kilowatt-hours. An electricity meter records – the energy transferred into a house.

2 Power = 0.80 kW and time is 1.5 h, so energy used is 0.8 kW × 1.5 h = 1.2 kWh; the cost is 1.2 kWh × 15 p/kWh = 18 p

3 a How much of the electrical energy entering the TV is transferred to various other sorts of energy

  b Light 10 J

4 An arrow splits – where there is an energy transfer. Electrical energy flows – in from the left. Waste heat energy flows – out downwards. The electrical energy input is – equal to the sum of the energy outputs. The thickness of each arrow is – proportional to its energy. Useful transferred energy flows – out to the right

5 Efficiency = 20/50 = 0.4 or 40%

6 Put very expensive taxes on inefficient cars; subsidise house insulation; replace old power stations with more efficient ones

7 The world's population is increasing; so each of those extra people will need a certain amount energy. At the moment, a lot of the energy is used by the few people in rich countries; and it is likely that everyone else will also want to have the same energy-rich lifestyle

Page 99 Generating electricity

1 Generators; magnet; coil

2 Voltage

3 e, a, d, b, c

4 a Coal; gas; nuclear fuel; oil; biofuels such as wood *(Any 3)*

  b Wind; water behind dams; tidal

5 It spins; the magnet inside the generator

Answers

6 Exhaust from burning gas; the air as wind; and water in rivers and the sea

7 The fuel is burnt; and transfers energy to water; which boils to make steam; which passes through the turbine

8 The gas is burnt; to make a high pressure gas; which passes through the first turbine. When the gas leaves the turbine it is hot enough to boil water into steam; which is sent through the second turbine

9 The rods of uranium; fuel are stacked close together; in the reactor. Nuclear reactions transfer energy in the rods, heating them up. This thermal energy is carried away by high pressure water; pumped through the stack of rods to a heat exchanger where it is used to boil water to make steam; for the turbine

Page 100 Electricity matters

1 It is radioactive; so emits radiation

2 a The radiation from nuclear waste is ionising; so it damages cells in your body

  b Stay well away from it; or put thick shields of metal and concrete in the way

  c When you are contaminated some of the waste has got inside your body. This means that you are exposed to the radiation for a much longer time; and it is difficult to isolate and remove the material

3 They compile statistics of deaths caused by each technology; taking into account the amount of electricity produced and how many years it has been used. They can then decide which technology is low risk (smaller number of deaths per MWh of electricity)

4 Hydroelectric; tidal; wind

5 Hydroelectricity can be switched on and off quickly, so is useful for meeting surges in demand; and it can be used to store excess electricity from elsewhere by pumping water back behind the dam. However, the dams cost a lot to build and the lakes behind them flood a lot of land

6 You can put wind turbines out at sea; so that they can't visually pollute the landscape. You can use the dams of hydroelectric schemes; to prevent the sudden flooding of rivers downstream when there is a lot of rain

7 230 V

8 The most important wasteful energy transfers occur in the power station. Whenever the energy is being transferred as heat, some of it escapes into the environment. Some energy has to be used to provide current for the electromagnet in the generator; and finally some is lost as heat in the cables of the National Grid

Page 101 Electricity choices

1 Fossil fuel – produces greenhouse gas; Wind power – noise and visual pollution; Nuclear power – produces radioactive waste; Hydroelectricity – floods large areas of land

2 Coal; oil; gas; nuclear

3 Geothermal; nuclear; solar; wind

4 They could install 40 wind farms, but this would take up a lot of land and upset a lot of people who think they are ugly. Some could go out at sea where they would work better and not be seen. They could build 20 nuclear power stations, but that would mean a lot of radioactive waste to deal with in the future. I think they should go for a mixture of both; so that they don't have to rely on just one technology

5 People in the USA and the UK will both be affected by the global warming caused by the carbon dioxide

6 There will be more people on Earth in future which means less energy per person to keep the total energy the same. Most people around today use less energy than we do in the UK, so future people will need more energy to adopt our lifestyle. So we will need to use a lot less energy in the future

7 They need to make sure that they can always generate enough electricity to meet demand. To avoid over-dependence on just one energy source, a variety of technologies should be used to generate electricity. Since fossil fuels will inevitably run out, they need to invest in a range of renewable energy sources. They could also make sure that old power stations are replaced by more efficient ones to make better use of the energy source

Page 102 P3 Extended response question

5–6 marks
Answer includes nearly all of the relevant points, as follows:
- Is the cost of the energy source low enough?
- Is the energy source near enough to be reliable?
- Does the energy source contribute to global warming?
- How long will the energy source last for?
- Will the power station be efficient?
- How easy is it to dispose of waste from the power station?
- How much it will cost to run and maintain the power station?
- What is the public's perception of risks from the power station?

All information in answer is relevant, clear, organised and presented in a structured and coherent format. Specialist terms are used appropriately. There are few, if any, errors in grammar, punctuation and spelling

3–4 marks
Answer includes at least half of the relevant points (see above). For the most part the information is relevant and presented in a structured and coherent format. Specialist terms are used for the most part appropriately. There are occasional errors in grammar, punctuation and spelling

1–2 marks
Answer includes at least two relevant points (see above). Answer may be simplistic. There may be limited use of specialist terms. Errors of grammar, punctuation and spelling prevent communication of the science

0 marks
Insufficient or irrelevant science. Answer not worthy of credit

P4 Explaining motion

Page 103 Speed

1 a Speed = distance ÷ time = 60 ÷ 3 = 20 km/h

  b Speed = distance ÷ time = 60000 ÷ (3 × 3600) = 5.5 m/s

2 Time = distance ÷ speed = 100 ÷ 20 = 5 hours

3 The average velocity will be less than 150 km/h; because the displacement of the journey will be lower than the total distance; the railway line will not be a straight line

4 a Constant high speed   b Speeding up

  c Stationary   d Constant slow speed

5 a 1000 m

  b Speed = 1000 m ÷ (10 × 60) = 0.17 m/s

  c She stopped for 15 minutes; walked back 400 m in the next 5 minutes; stopped for 5 minutes; then went back 600 m in the next 15 minutes

  d 2000 m

  e 0 m

  f Average speed = 2000 ÷ (50 × 60) = 0.67 m/s; average velocity = 0 m/s; average velocity is zero because she went back to the original starting point

Page 104 Acceleration

1 Car B

2 100 km/h = 100 000 m ÷ 3600 s = 27.8 m/s; Car A's acceleration = change in speed ÷ time = 27.8 ÷ 12 = 2.3 m/s²; Car B's acceleration = 27.8 ÷ 10 = 2.8 m/s²

3 Change in speed = acceleration × time = 10 × 1.6 = 16 m/s. Answer = 16 m/s

4 Speed increases steadily; for first 20 seconds; up to maximum speed of 10 m/s; then remains at this constant speed for 40 seconds

5 Graph needs to show: a straight line from the origin to 8 m/s at 20 s; a horizontal line at 8 m/s from 20 to 40 seconds; a straight line from (40, 8) to (50, 5); a horizontal line at 5 m/s from 50 to 60 seconds

6 a All points correctly plotted (3 marks); continuous line drawn between points (1 mark)

  b Acceleration = gradient of line = 12 ÷ 20 = 0.6 m/s²

  c Deceleration = (16 - 4) ÷ 15 = 0.8 m/s² (ignore negative sign)

Page 105 Forces

1 Pairs; a downwards; an upwards

2 a The ball pushes on the tennis racket backwards

  b The cart pulls the horse backwards

  c The Earth is attracted to the Moon

3 Version B

Table pushes up on book

Weight of book pushes down on table

4 a To make the results more reliable

  b 32.6

  c Same block of wood

  d Average values 9.7; 25.5 or 27.8; 16.5 (1 mark each)

  e Glass; it is much smoother than wood and carpet (1 mark each)

5 Useful friction on brakes/tyres; friction is used to slow the wheels down/friction is needed to grip the road; there is unwanted friction on axles/gears/chain, etc.; friction on moving parts causes heat/wear/energy loss

Page 106 Effects of forces

1 a 50 N downwards   b 50 N upwards

   c 100 N downwards   d 100 N upwards

2 a Diagram needs to include: downwards arrow labelled weight/gravity; upwards arrow labelled air resistance/drag

   b Downwards force must be bigger than upwards force

   c Speed eventually becomes constant; when drag force is equal in size to weight

3 b and c

4 a Momentum = $0.3 \times 30 = 9$ kg m/s

   b Momentum = $1100 \times 15 = 16\ 500$ kg m/s

   c Momentum = $4000 \times 5 = 20\ 000$ kg m/s

5 The ball has a lot of momentum; when he catches it the momentum drops to zero. If he moves his hands towards his body as he catches it the time to stop the ball increases; this reduces the force

Page 107 Work and energy

1 Destroyed; transferred; conservation

2 a, c

3 a Work done = force × distance = $3 \times 1.3 = 3.9$ J

   b Gravitational potential energy

   c 3.9 J

   d KE = $\frac{1}{2}$ mv^2, so v^2 = 2 × KE ÷ m; v^2 = 2 × 3.9 ÷ 0.3 = 26; v = $\sqrt{26}$ = 5.1 m/s

4 a The diagram should show: maximum kinetic energy at the lowest point; maximum potential energy at either high point

   b GPE = weight × height = $1 \times 0.2 = 0.2$ J

   c KE = $\frac{1}{2}$ × mass × v^2 = $\frac{1}{2}$ × 0.1 × 1.2^2 = 0.072 J

   d GPE = 0.2 − 0.072 = 0.128 J

5 a The roller coaster starts with gravitational potential energy (GPE); which is converted to kinetic energy (KE) on the way down the track. As it goes up again the KE is converted back to GPE again

   b Some of the energy is dissipated as heat; due to friction and air resistance; so it does not have as much GPE as it started with

Page 108 P4 Extended response question

Relevant points include:

• safety devices that protect in the event of a crash: crumple zones; seat belts; air bags

• how they work: increase time taken for person to slow down; slowing down momentum change; reducing force on the person

5–6 marks

Mentions all three devices (see above). Clearly links reduction in force during a collision to the increase in time needed to change momentum. All information in answer is relevant, clear, organised and presented in a structured and coherent format. Specialist terms are used appropriately. There are few, if any, errors in grammar, punctuation and spelling

3–4 marks

Mentions at least two devices that protect passengers during a crash (see above). Includes two out of three points about how they work. For the most part the information is relevant and presented in a structured and coherent format. Specialist terms are used for the most part appropriately. There are occasional errors in grammar, punctuation and spelling

1–2 marks

Mentions at least two devices that protect passengers during a crash (see above). Includes one relevant point about how they work. Answer may be simplistic. There may be limited use of specialist terms. Errors of grammar, punctuation and spelling prevent communication of the science

0 marks

Insufficient or irrelevant science. Answer not worthy of credit

P5 Electric circuits

Page 109 Electric current – a flow of what?

1 Inside nucleus – proton – positive; Inside nucleus – neutron - neutral; Orbiting nucleus – electron – negative

2 Electron; negative; attracted

3 Metals are conductors; they contain free electrons

4 The movement/friction of the iron charges up the shirt; electrons are transferred between Sadie/the iron and the shirt. Sadie becomes charged up. When she touches the wall she discharges/electrons move away. An electric current causes the small shock

5 2 p coin; steel scissors

6 The cell gives energy to the electrons/charged particles; the electrons are free to move in the connecting wires; electrons/charged particles carry the energy to the bulb; at the bulb the energy is transferred to light; and heat

7 a Arrow showing flow of charged particles towards positive terminal of cell

   b Electrons are negatively charged; so they will be attracted towards the positive terminal

   c Put another cell (in series with the first one); so more energy will be supplied to each electron

Page 110 Current, voltage and resistance

1 a Circuit diagram should be amended to show correct voltmeter symbol; correct ammeter symbol; voltmeter placed in parallel to either cell or bulb; ammeter placed in series (either side of the bulb)

   b The charged particles carry the energy from the cell to the bulb; the charged particles in this circuit are electrons; the electrons/charged particles are given energy in the cell; the electrons/charged particles flow all the way round the circuit; at the bulb the energy carried by the electrons/charged particles is converted to heat and light (Any 3)

2 Potential difference is the scientific term for voltage and is measured in volts; it is a measure of how much energy is given to/converted from the charged particles; 1 volt means that 1 joule of energy is given to/converted for each unit of charge

3 B, A, C, D

4 Resistance = voltage ÷ current = 12 ÷ 0.5 = 24 Ω

5 a All points correctly plotted; line of best fit drawn accurately

   b Either: calculate gradient of line; then calculate 1/gradient, Or: calculate a value for resistance from table values; calculate two or more values for resistance and work out average (value for resistance should be in the range 16 Ω < R < 18 Ω)

   c He should repeat the experiment; and work out average values of current at each voltage

Page 111 Useful circuits

1 a Current decreases

   b Current increases

   c Current would be zero

2 a The series circuit would have dimmer bulbs

   b Series circuit – all bulbs would go out; parallel circuit – all the other bulbs would stay on with the same brightness

3 A$_1$ = 0.9 A; A$_2$ = 0.6 A; A$_3$ = 0.3 A; A$_4$ = 0.9 A

4 The resistance of an LDR varies with light intensity; the resistance of a thermistor varies with temperature; Either: the lower the light intensity the higher the resistance (or vice versa), Or: the lower the temperature the higher the resistance (or vice versa)

5 He records the resistance at a known temperature; he records the resistance at a second known temperature; the lower known temperature could be melting ice; the higher known temperature could be boiling water; he makes a scale between the two known temperatures; it is not a linear temperature scale; record values of resistance at several different known temperatures and plot a graph; read the temperature off the graph (Any 4)

6 a With two components you add the resistances; twice as much resistance would give half the current

   b Bulbs do not have constant resistance; at lower currents the bulbs are cooler; so they have lower resistance; so more current will flow

Page 112 Producing electricity

1 Field; magnetic; iron; compasses; poles

2 a The output voltage would be higher

   b The voltage produced would be in the opposite direction

   c The output voltage would become very low

3 a Torch – d.c.   b Transformer – a.c.

   c Computer – d.c.   d Iron – a.c.

4 Power = V × I = 230 × 3.5; = 805; W

5 a Current = power ÷ voltage = 100 ÷ 230; = 0.43; A

   b New current = 0.43 ÷ 4 = 0.11; power = 230 × 0.11 = 25 W

Page 113 Electric motors and transformers

1 Magnetic field lines drawn similar to those shown; arrows on field lines from South towards North marked on diagram

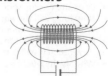

Answers

2 The force on the sides of the coil/wire carrying current is at right angles to both the magnetic field; and to the current. The forces on either side of the coil produce a turning effect; once every half turn the current in the coil has to change direction; in order for the turning effect to be in the same direction

3 Voltage; alternating; coils; iron; primary; secondary

4 i a Step-down b Step up c Step-down d Neither

ii V_1, V_3, V_4, V_2

5 The output from a battery is d.c.; d.c. will create a constant magnetic field; for a voltage to be induced on secondary coil; there must be a varying magnetic field

Page 114 P5 Extended response question

5–6 marks
Includes all main details and some additional details. Main details:
- rotate magnet
- to alter magnetism/magnetic field of iron/coil
- voltage across/current in coil.

Additional details:
- process is called (electromagnetic) induction
- voltage keeps on changing / a.c. / not d.c.
- current in components connected to ends of coil
- work done turning magnet transfers to electrical energy.

All information in answer is relevant, clear, organised and presented in a structured and coherent format. Specialist terms are used appropriately. There are few, if any, errors in grammar, punctuation and spelling

3–4 marks
Includes some of the main details and some additional details (see above). For the most part the information is relevant and presented in a structured and coherent format. Specialist terms are used for the most part appropriately. There are occasional errors in grammar, punctuation and spelling

1–2 marks
Includes at least one main detail and at least one additional detail (see above). Answer may be simplistic. There may be limited use of specialist terms. Errors of grammar, punctuation and spelling prevent communication of the science

0 marks
Insufficient or irrelevant science. Answer not worthy of credit

P6 Radioactive materials

Page 115 Nuclear radiation

1 Nucleus; negatively charged; protons; electrons; mass

2 a to iii; b to i; c to ii

3 An atom of the same element; with different number of neutrons in nucleus; and the same number of protons

4 An atom is neutral / has same number of protons and electrons; an ion has an overall charge; less protons than electrons / more protons than electrons

5 a Background radiation is ionising radiation which is all around us; from cosmic rays / rocks / man-made sources

   b Some rocks are more radioactive than others

6 Michael is correct; radioactivity is not affected by physical conditions; it only depends on the amount of radioactive material present in the sample

Page 116 Types of radiation and hazards

1 Gamma; alpha; alpha; gamma; beta

2 Put the different materials in between the mantle and the detector. Find out if the reading on the detector changes; if the radiation is stopped by paper it is alpha; if it is stopped by aluminium or steel it is beta; if it is not stopped by any of the materials it is gamma

3 A helium nucleus; with two protons and two neutrons; a large, positively charged particle

4 a 900; microSieverts

   b The bar from nuclear power and weapons would be much higher; because a lot more radiation was released as the power station was damaged

5 Ionising radiation ionises (water) molecules in the body; the ions can then react with other molecules in the body; such as DNA

Page 117 Radioactive decay and half-life

1 Once a radioactive nucleus (atom) has emitted some radiation; it can't do it again; so the number of radioactive nuclei decreases

2 Alpha and beta particles are made up from parts of the nucleus; so the number of protons in the nucleus changes; gamma rays are electromagnetic waves carrying energy and do not change the nucleus's structure

3 4; 2; He; 212; 82

---

4 a A   b C   c C

5 a All points plotted correctly; smooth curve drawn (1 mark for each)

   b One value of half-life read off; average of more half-life values calculated (1 mark for each)

Page 118 Uses of ionising radiation and safety

1 a Gamma

   b The instruments can be sealed in packaging; and the radiation will pass through the packaging

2 a So radiation is concentrated on the cancer tissue; and is not going to damage healthy cells around the cancer

   b Beta radiation will just pass though skin; and won't damage deeper tissue

3 a It will get less

   b Beta

   c Alpha will not get through any thickness of paper; gamma passes through the paper too easily, so too little variation as paper changes thickness

4 b

5 Health/cancer risk for all participants due to irradiation by the rod; this risk is greatest for the radiographer who will repeat the procedure many times; patient will benefit if their existing cancer is cured; but the risk of patient and/or radiographer developing a new cancer may outweigh the benefits of the procedure (Any 3)

Page 119 Nuclear power

1 Fuel; fission; fusion

2 $E = 0.24 \times (3 \times 10^8)^2; = 2.16 \times 10^{16}$ J

3 Advantages: no CO_2 given off; no contribution to greenhouse effect; large amounts of energy from small amounts of fuel; fuel is not running out (Any 2)

Disadvantages: employees at risk of radiation; radioactive waste produced; risk of nuclear accident (Any 2)

4 a The neutron is absorbed by the uranium nucleus, making it unstable; it then splits into two smaller nuclei; and gives off two or three neutrons; which then go on to hit more uranium nuclei causing them to fission

   b Control rods (boron); are used to absorb some of the neutrons

5 Helium

6 a Larger quantities of energy could be available; ready source of fuel/hydrogen; no radioactive waste (Any 2)

   b The fuel needs to be contained; at very high temperatures and pressures. This uses up more energy than is created from the reaction

Page 120 P6 Extended response question

Relevant points include:
- high level only produced in reactor; is very radioactive; so is stored in ponds of water; until it becomes intermediate waste / less radioactive
- hospital produces mostly intermediate waste; intermediate waste is encased in concrete / glass; and stored in metal drums; under guard / in secure conditions
- low level waste is produced at both hospital and reactor; is put in landfill; with waterproof linings; to keep radioactivity out of ground water
- all radioactive waste is harmful / cancerous; becoming less harmful as time goes on

5–6 marks
Evaluates production and use of the radioactive materials, and correctly identifies sources for all three types of waste. Suggests how to dispose of them safely. Gives a valid reason why waste needs to be stored carefully. All information in answer is relevant, clear, organised and presented in a structured and coherent format. Specialist terms are used appropriately. There are few, if any, errors in grammar, punctuation and spelling

3–4 marks
Evaluates production and/or use of the radioactive materials, and correctly identifies sources for at least two types of waste, perhaps omitting some important details. For the most part the information is relevant and presented in a structured and coherent format. Specialist terms are used for the most part appropriately. There are occasional errors in grammar, punctuation and spelling

1–2 marks
Refers to at least one type of waste and valid disposal method for it; may not give a reason for the need for careful disposal. Answer may be simplistic. There may be limited use of specialist terms. Errors of grammar, punctuation and spelling prevent communication of the science

0 marks
Insufficient or irrelevant science. Answer not worthy of credit

Answers

P7 Further physics – studying the Universe

Page 121 The solar day

1 East–west

2 a 24 hours

  b The time taken for the Sun to appear at the same place in the sky from one day to the next

3 stars; orbit; Earth; east; west

4 The Earth rotates a full 360 degrees on its axis in 24 hours; this makes the Sun and stars appear to move across the sky once every 24 hours

5 a 3

  b 2

  c 1

6 23 hours 56 minutes

7 As the Earth rotates on its axis it is also orbiting the Sun; a sidereal day is the time taken to make one complete revolution; because the Earth is also orbiting the Sun the Earth needs to rotate a few degrees more before the Sun is in the same position

Page 122 The Moon

1 a T

  b T

  c F

  d F

2 New Moon

3 a 4

  b 3

  c 1

  d 2

4 A lunar eclipse is the Earth's shadow on the Moon, a solar eclipse is the Moon's shadow on the Earth; a lunar eclipse lasts longer than a solar eclipse; a lunar eclipse covers a wider area than a solar eclipse

5 The Earth, Sun and Moon do not orbit on exactly the same plane; eclipses only occur when they line up directly

Page 123 The problem of the planets

1 Their size; how close they are to Earth; how reflective their surface is

2 The rotation of the Earth on its axis

3 The planets appear to change speed; the planets sometimes appear to move backwards

4 Retrograde motion

5 As they orbit the Sun at a different speed to the Earth; sometimes planets are closer to Earth and sometimes further away

6 T, T, F, F

7 The Earth orbits at a faster speed than Mars; when compared to the background of distant stars; as the Earth approaches Mars, Mars appears to move towards Earth; once the Earth has passed Mars, Mars appears to move away from Earth

Page 124 Navigating the sky

1 The night sky is the side that points away from the Sun; in different months of the year this side points in different directions

2 I would expect more stars to be visible in July than in January

3 a Proxima Centauri

  b Polaris

  c Barnard's Star; Altair

4 The apparent path of the Sun around the Earth across the year

5 The angle of ascension is measured from the point that the ecliptic crosses to the Northern hemisphere

6 The angle formed by the equator and position of the star directly above or below the equator

7 Because the Earth rotates once in 24 hours, 15 degrees rotation is the same as one hour

Page 125 Refraction of light

1 Its speed changes

2 It is slower in glass

3 Refraction

4 When crossing the boundary waves turn inwards; wavelength becomes shorter

5 Ray entering at right angles passes straight through with refracting; ray entering at angle, turns in when entering block; and turns out again when exiting

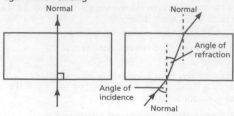

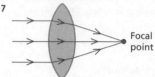

6 It is fatter in the middle than at the edges

7

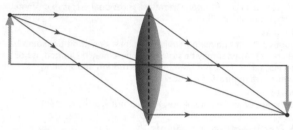

Focal point

Page 126 Forming an image

1 The distance from the centre of the lens to the point where parallel rays are brought to a focus

2 The curvature of the lens; the material of the lens

3 a 1/0.05 = 20 Dioptres

  b 1/2 = 0.5 metres

  c Lens A

4 distant; parallel; focal point; principal axis

5

6 Extended objects`

Page 127 The telescope

1 F, T, T, F

2 It means the image appears to be three times larger; than if observed with the naked eye

3 a 200/10 = 20 times

  b 90/5 = 18 times

4 a 50 dioptres

  b 20 times

  c 0.4 dioptres

Page 128 The reflecting telescope

1 They can be made larger; they are lighter in weight; they are easier to fix in a telescope; they do not suffer from chromatic aberration (*Any 2*)

2 The objective mirror is the largest curved one at the back of the telescope; the eyepiece lens is off to the side

3 c, a, d, b

4 White light contains a mix of colours; different colours are refracted differently; like a prism this separates the colours as light passes through the lens

5 Objects do not appear the correct colour / images have coloured edges

6 Mirrors use reflection to focus the light rather than refraction

Page 129 Diffraction

1 To see faint objects; to give sharp images / reduce diffraction

2 The amount of diffraction would increase

3 Light has a much shorter wavelength than the width of the doorway

4 Diffraction occurs more the longer the wavelength; radio waves have a long wavelength so are diffracted into the valley; mobile phone signals have a short wavelength so are not diffracted as much; and are therefore blocked by the hills

5 a C, B, A

  b The smaller the aperture the greater the diffraction; light passing through a large aperture is diffracted the least and gives crisp images; light entering the narrow aperture is diffracted more and gives a fuzzy image

6 Radio waves have a long wavelength; so need a wide aperture to minimise diffraction

Answers

Page 130 Spectra

1 Its temperature; and its composition

2 Different colours of light are refracted different amounts; the shape of the prism means that white light is split entering the prism and split further exiting the prism

3

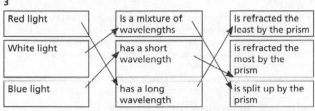

Red light		Is a mixture of wavelengths		Is refracted the least by the prism
White light		has a short wavelength		is refracted the most by the prism
Blue light		has a long wavelength		is split up by the prism

4 The gaps need to be about the same wavelength as the light; which has a very short wavelength

5 The spacing of the grooves on a DVD is different to a CD

6 Other quasars had not been discovered so there was not much evidence; astronomers did not believe something so far away could be visible; the findings and conclusions needed to be peer reviewed *(Any 2)*

Page 131 The distance to the stars

1 Because a dim star could be a long way away or simply a low power star and vice versa

2 Distant stars

3 July

4

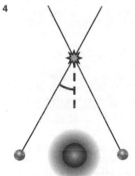

5 Sirius = 0.38 arc seconds; Arcturus = 22.7 pc; Barnard's Star = 1.82 pc; Proxima Centauri – 0.77 arc seconds

6 Distance in parsecs = 1/parallax angle in arcseconds
Distance = 1/0.5 = 2 parsecs

7 Because the parallax angle would be too small to measure

Page 132 Distance and brightness in the Universe

1 The distance travelled by light in one year

2 light-years; parsecs; mega parsecs

3 T, F, T, F

4 Luminosity is how much energy is given out, apparent brightness is how much reaches Earth

5 The inverse square law means, for example, that doubling one thing quarters the other

6 a 400 lumens
  b 25 lumens
  c 1 lumen

7 The large blue star must be further away; it is blue so hotter; and larger so gives off more radiation

Page 133 Cepheid variables

1 A star that varies in brightness; with a regular frequency / period

2 The greater period, the more luminous

3 T, F, T

4 The one with a period of 10 days is closest; it has a lower period so lower luminosity; so must be closer to appear to be the same brightness

5 Cepheid variables allowed astronomers to work out the distance to galaxies; that were too far away for other methods to work

Page 134 Galaxies

1 Galaxy; the Milky Way; the Sun; stars; the Universe

2 Nebula

3 a This hypothesis said that nebulae were galaxies like our own but a very long way away
  b These astronomers thought they were close-up clouds of dust and gas

4

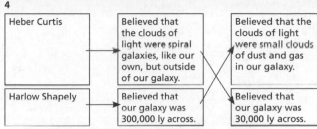

| Heber Curtis | | Believed that the clouds of light were spiral galaxies, like our own, but outside of our galaxy. | | Believed that the clouds of light were small clouds of dust and gas in our galaxy. |
| Harlow Shapely | | Believed that our galaxy was 300,000 ly across. | | Believed that our galaxy was 30,000 ly across. |

5 There was not enough evidence to prove either argument conclusively; as the evidence could be interpreted in different ways

Page 135 The expanding Universe

1 Around 14 billion light years

2 He looked at its period to work out its luminosity; he then compared the luminosity to its apparent brightness

3 a That they are moving away from us
  b Because the further away they are, the faster they are moving
  c The Big Bang Theory

4 As telescopes and technology developed, methods for accurately making measurements of long distances got better

5 NGC4414 – 1400 km/s
Whirlpool = 7 Mpc

6 a Because different galaxies are moving at different speeds and measurements of such long distances are not 100% accurate; by using many measurements an average can be found
  b –0.001/0.8 – 0.00125 (km/s)/Mpc

Page 136 The radiation from stars

1 The hotter the object the more radiation it emits; and the higher the peak frequency it emits

2 F, T, T, F

3 The luminosity of a star depends on its size and its temperature; it must be extremely large but cooler *(Also accept it must be a red supergiant)*

4 The core

5 The photosphere

6 Core ⟶ radiation zone ⟶ convection zone ⟶
photosphere ⟶ space
(Lose 1 mark for each one in the wrong place)

Page 137 Analysing stellar spectra

1 A line spectrum has radiation at just a few frequencies, but a continuous spectrum covers a full range

2 A line spectra / just radiation at a few discrete wavelengths

3 You would see a continuous spectra; with dark lines showing an absorption spectrum exactly matching the emission spectrum of the gas

4 Its composition

5 There will be more dark lines in the larger star; because it has a greater range of elements

6 Electrons, energy, energy, photons, light, elements, elements

7 Elements in the outer atmosphere of stars; absorb radiation at specific frequencies; studying what frequencies are absorbed indicates what elements are present

Page 138 Absolute zero

1 a B, C, A, D
  b A, C, B, D

2 They get faster

3 The kinetic energy of the molecules

4 –273 °C

5 Kelvins

6 As the gas is heated the particles move faster; so hit the end of the piston harder; pushing the piston out and making the volume increase / gas expand

7 As it is cooled the particles move slower; as they move slower the pressure falls; at 150 K the pressure will be 0.5 atm; once cooled to absolute zero the particles stop moving and the pressure is zero

Page 139 The gas laws

1 a The molecules in the balloon move around randomly; they collide with the inside surface walls of the balloon; exerting a force / pressure that gives the balloon its shape
  b The pressure would increase

2 If the temperature is kept constant the pressure is proportional to the volume

Answers

3 a 6 atmospheres

b 12 m³

4 a T1 = 273 + 20, T2 = 273 + 40;

$$\frac{P1}{T1} = \frac{P2}{T2}, \frac{100}{293} = \frac{P2}{313},$$

$$P2 = \frac{(100 \times 313)}{293} = 106.8 \text{ atm}$$

b $\frac{P1}{T1} = \frac{P2}{T2}. \frac{100}{293} = \frac{90}{T2}$;

$$T2 = \frac{(100 \times 293)}{100} = 263.7;$$

263.7 −273 = −9.3 °C

c P1 × V1 = P2 × V2, 100 × 5000 = 1 × V2;
V2 = 500,000 cm³

d $\frac{P1}{T1} = \frac{P2}{T2}, \frac{100}{293} = \frac{200}{T2}$;
T2 = (200 × 293)/100. T2 = 586;
It would rupture once above 313 °C

Page 140 Star birth

1 hydrogen and helium

2 As gravity pulls them together the kinetic energy increases; and the pressure increases

3 Density, hotter, pressure, temperature

4 Nuclear fusion

5 F, T, T, F

6 So that the density; and pressure; increase enough so the temperature is hot enough for fusion to begin

Page 141 Nuclear fusion

1 The core

2 Because the temperature is so high

3 Energy

4 An isotope of an element has the same number of protons but different numbers of neutrons

5 a 1

b 3

c 2

6 Mass number and atomic number on either side of the equation need to balance; the positron is released to ensure charge is conserved

7 X = 3

Y = 13

8 E = mc²

9 4 (1.674 × 10⁻²⁷ kg.) − 1 (6.645 × 10⁻²⁷ kg.) = 0.051 × 10⁻²⁷ kg
E = (0.051 × 10⁻²⁷ kg) (3 × 10⁸ m/s)² = 4.3 × 10⁻¹² Joules

Page 142 The lives of stars

1 Gravity pulling inwards; and pressure pushing outwards

2 When it runs out of hydrogen / begins to run out of hydrogen

3 The sun is not hot enough or at high enough pressure; but larger stars are hotter and denser

4 Iron

5 C, E, F

6 There is no fusion taking place in white dwarf stars

Page 143 The death of a star

1 B, A, C, E, D *(Lose one mark for each incorrectly placed)*

2 a hydrogen, helium

b copper, zinc

c The red supergiant is hotter and denser than the Sun

3

Sun-sized star	Expands to become a red giant or red supergiant.	Explodes in a supernova then contracts to become a neutron star.
Larger star	Expands to become a red giant.	Explodes in a supernova the contracts to become a black hole.
Largest stars	Expands to become a red supergiant.	Fusion stops, becomes a white dwarf and eventually black dwarf.

4 Iron fusion does not produce energy; so with no more energy produced the pressure drops and gravity crushes the star

Page 144 The possibility of extraterrestrial life

1 They are very far away; light reflected from them is lost in the glare from their star; they don't emit light of their own

2 a Star B

b The observed brightness dips at regular intervals; as the orbiting planet passes in front of the star

3 The Goldilocks zone is a place around a star that is not too hot or cold; to have liquid water; it is important because all life as we know it requires liquid water

4 Could choose from Mars or Europa; evidence suggests that Mars had liquid water in the past, Europa may have liquid water now

Page 145 Observing the Universe

1 A, C, D

2 a High altitude; low atmospheric pollution; clear skies; low light pollution; low humidity *(Any 2)*

b Build cost; environmental impact; impact on local people; travel and working conditions of staff *(Any 2)*

c The air around the UK is quite humid; too cloudy; snowy winter weather; too cold for staff in winter; not particularly high enough *(Any 2)*

3 Remote access so reduce travel time and cost; more precise positioning; can be programmed to follow objects; can record to computer for automatic processing *(Any 3)*

Page 146 International astronomy

1 More traffic; more jobs; better infrastructure; more money spent in the area; possible environmental concerns; disruption during construction *(Any 2)*

2 a Shared cost; greater expertise

b Who gets to use it when

3 Any suitable example, e.g. searching for gamma ray bursts – Swift satellite, looks for the burst of gamma rays, sends message around the world. Other observatories home in on the area to try and catch the after-effects of the burst
Or

Supernova – neutrino observatories will detect first rush of neutrinos, send message across the globe to look for where they came on, when found tell others so all can study

4 Travel to the telescope; remote control over the internet; instruct the observatory staff what to do

5 Any suitable example, e.g. helping to study photographs that have been made available online that professionals do not have the time to look at
Or

Seti@home and similar crowd sourcing computer analysis programs

Page 147 Extended response question

5–6
The large aperture of a radio telescope reduces diffraction and allows it to detect very weak signals although some radio waves have such a long wavelength that they are still diffracted, which results in poor resolution images

Unlike optical telescopes radio telescopes can even detect objects that are too cool to emit visible light, however they can only detect radio waves so are limited to a small section of the electromagnetic spectrum

Radio telescopes are very large and expensive to build and some people consider them to be ugly but the construction and running of such a large observatory can bring additional jobs and wealth to the area, which benefits local people

3–4
A radio telescope has a very large aperture making it capable of detecting weak signals and reducing diffraction. However it can only detect radio waves so is limited to just a small section of the electromagnetic spectrum and some radio waves have such a long wavelength that they are still diffracted, causing fuzzy images

1–2 marks
The advantage of a radio telescope is that it is very big, which reduces diffraction, and being so big it can detect weak signals

0 marks
Insufficient or irrelevant science. Answer not worthy of credit